Alfa Romeo
BERLINAS
(SALOONS/SEDANS)

Other Veloce publications -

Colour Family Album Series
Alfa Romeo by Andrea & David Sparrow
Bubblecars & Microcars by Andrea & David Sparrow
Bubblecars & Microcars, More by Andrea & David Sparrow
Citroën 2CV by Andrea & David Sparrow
Citroën DS by Andrea & David Sparrow
Fiat & Abarth 500 & 600 by Andrea & David Sparrow
Lambretta by Andrea & David Sparrow
Mini & Mini Cooper by Andrea & David Sparrow
Motor Scooters by Andrea & David Sparrow
Porsche by Andrea & David Sparrow
Triumph Sportscars by Andrea & David Sparrow
Vespa by Andrea & David Sparrow
VW Beetle by Andrea & David Sparrow
VW Beetle/Bug, Custom by Andrea & David Sparrow
VW Bus, Camper, Van & Pick-up by Andrea & David Sparrow

SpeedPro Series
How to Blueprint & Build a 4-Cylinder Engine Short Block for High Performance by Des Hammill
How to Build a V8 Engine Short Block for High Performance by Des Hammill
How to Build & Modify Sportscar/Kitcar Suspension & Brakes for High Performance by Des Hammill
How to Build & Power Tune Weber DCOE & Dellorto DHLA Carburetors Second Edition by Des Hammill
How to Build & Modify SU Carburettors for High Performance by Des Hammill
How to Build & Power Tune Harley-Davidson 1340 Evolution Engines by Des Hammill
How to Build & Power Tune Distributor-type Ignition Systems by Des Hammill
How to Build, Modify & Power Tune Cylinder Heads Second Edition by Peter Burgess
How to Choose & Time Camshafts for Maximum Power by Des Hammill
How to Give Your MGB V8 Power Updated & Revised Edition by Roger Williams
How to Plan & Build a Fast Road Car by Daniel Stapleton
How to Power Tune BMC/BL/Rover 998cc A Series Engines by Des Hammill
How to Power Tune BMC/BL/Rover 1275cc A Series Engines by Des Hammill
How to Power Tune the MGB 4-Cylinder Engine by Peter Burgess
How to Power Tune Alfa Romeo Twin Cam Engines by Jim Kartalamakis
How to Power Tune Ford SOHC 'Pinto' & Sierra Cosworth DOHC Engines by Des Hammill
How to Power Tune Midget & Sprite for Road & Track Updated, Revised & Enlarged Edition by Daniel Stapleton
How to Improve MGB, MGC, MGB V8 by Roger Williams

General
Automotive Mascots: A Collectors Guide to British Marque, Corporate & Accessory Mascots by David Kay & Lynda Springate
Bentley Continental Corniche & Azure 1951-1998 by Martin Bennett
Alfa Romeo Giulia Coupe GT & GTA by John Tipler
British Cars, The Complete Catalogue of 1895-1975 by Culshaw & Horrobin
British Trailer Caravans & their Manufacturers 1919-1959 by Andrew Jenkinson
British Trailer Caravans from 1960 by Andrew Jenkinson
Bugatti Type 40 by Barrie Price
Bugatti 46/50 Updated & Revised Edition by Barrie Price
Chrysler 300 - America's Most Powerful Car by Robert Ackerson
Cobra - The Real Thing! by Trevor Legate
Cortina- Ford's Best Seller by Graham Robson
Daimler SP250 'Dart' by Brian Long

Datsun Z - From Fairlady to 280Z by Brian Long
Datsun/Nissan 280ZX & 300ZX by Brian Long
Dune Buggy Handbook, A-Z of VW-based Buggies since 1964 by James Hale
Fiat & Abarth 124 Spider & Coupé by John Tipler
Fiat & Abarth 500 & 600 (revised edition) by Malcolm Bobbitt
Ford F100/F150 Pick-up by Robert Ackerson
Jim Redman - Six Times World Motorcycle Champion by Jim Redman
Grey Guide, The by Dave Thornton
Lea-Francis Story,The by Barrie Price
Lola - The Illustrated History (1957-1977) by John Starkey
Lola T70 - The Racing History & Individual Chassis Record New Edition by John Starkey
Lotus 49 by Michael Oliver
Mazda MX5/Miata 1.6 Enthusiast's Workshop Manual by Rod Grainger & Pete Shoemark
Mazda MX5/Miata 1.8 Enthusiast's Workshop Manual by Rod Grainger & Pete Shoemark
Mazda MX5 - Renaissance Sportscar by Brian Long
MGA by John Price Williams
Mini Cooper - The Real Thing! by John Tipler
Motor Museums of the British Isles & Repulic of Ireland by David Burke & Tom Price
Porsche 356 by Brian Long
Porsche 911R, RS & RSR New Edition by John Starkey
Porsche 914 & 914-6 by Brian Long
Prince & I, The (revised edition) by Princess Ceril Birabongse
Rolls-Royce Silver Shadow/Bentley T Series, Corniche & Camargue Updated & Revised Edition by Malcolm Bobbitt
Rolls-Royce Silver Spirit, Silver Spur & Bentley Mulsanne by Malcolm Bobbitt
Rolls-Royce Silver Wraith, Dawn & Cloud/Bentley MkVI, R & S Series by Martyn Nutland
Singer Story:Cars, Commercial Vehicles, Bicycles & Motorcycles by Kevin Atkinson
Taxi! The Story of the 'London' Taxicab by Malcolm Bobbitt
Triumph Motorcycles & The Meriden Factory (paperback) by Hughie Hancox
Triumph Tiger Cub by Mike Estall
Triumph TR6 by William Kimberley
Velocette Motorcycles - MSS to Thruxton by Rod Burris
Volkswagen Karmann Ghia by Malcolm Bobbitt
VW Bus, Camper, Van, Pickup by Malcolm Bobbitt
VWs of the World by Simon Glen
Works Rally Mechanic 1955-1979 by Brian Moylan

First published 2000 by Veloce Publishing Plc., 33, Trinity Street, Dorchester DT1 1TT, England.
Fax: 01305 268864/e-mail: veloce@veloce.co.uk/website: http://www.veloce.co.uk
ISBN: 1-901295-74-5/UPC: 36847-00174-2

Readers with ideas for automotive books, or books on other transport or related hobby subjects, are invited to write to Veloce Publishing at the above address.
British Library Cataloguing in Publication Data -
A catalogue record for this book is available from the British Library.
Typesetting (Bookman), design and page make-up all by Veloce on AppleMac.
Printed in Hong Kong.

Visit Veloce on the Web - www.veloce.co.uk

Alfa Romeo
BERLINAS

(SALOONS/SEDANS)

John Tipler

VELOCE PUBLISHING PLC

PUBLISHERS OF FINE AUTOMOTIVE BOOKS

Alfa Romeo Berlinas

THANKS

This book has been a long time coming, and recognition of Alfa Romeo Berlinas is well overdue. So I'd like to express my sincere thanks to the following people for sharing anecdotes about their cars and for lending photographs for the book.

Starting with the manufacturer, Mrs Elvira Ruocco at the Alfa Romeo *Centro Storico Documentazione* at Arese provided some of the archive material. Also, the staff at Alfa GB, including 'Josh' Puneet Joshi, Angie Voluti, Giulietta Calabrese and Peter Newton. Also helpful were Bob Hoare and Richard Gadeselli, who is Head of Corporate Affairs at Fiat UK.

As ever, AROC Secretary Michael Lindsay delved into his personal photo albums and came up with some fine shots of saloon cars in action at AROC meetings and races.

Other pictures, and Alfa Romeo sales brochures, came from Alfasud specialist and Auto Italia championship winner Ian Brookfield, who co-runs the parts specialist Just Suds. The Giulietta Berlina T.I. pictures came from John Brittan; Patrick Walter sent me photos of his 146 ti. My daughter Keri sought out Cumbrian locations for our 155 V6 photo-shoot. Susheela Evitts, Matt Letch and Zoë Tipler held the fort at National Alfa Day, Richard Everton provided shots of his Giulia T.I. Super and his regular Super, while James Castle posed and Michel Rochebuet lamented his meagre lunch.

My comrade from JPS promotions days, Ian Catt, provided shots of GTAs at Snetterton and Super at Spa.

One of the UK's most accomplished Alfa restoration specialists, Mike Spenceley of MGS Coachworks, gave me a thorough guided tour of the short-comings of Giulia Supers in the body-work department, and unstintingly gave up a day so that I could snap his immaculate Super. Mike also provided the bulk of the restoration pictures. Thanks to AROC Chairman Ed McDonough for tales of his Giulietta T.I. discovery and the racing Rodriguez brothers. Stuart Taylor, secretary of the 105 Series Register, sent a variety of shots of Giulia saloons.

Norwich Alfa specialist and Alfasud racer, Richard Drake, of Richard Drake Motors, put the kettle on, and Alastair Kerridge of the Norwich Alfa dealership Lindfield Italia Ltd (now under the Desira banner) lent me a number of new Alfa demonstrators for evaluation and photo opportunities.

Roger Monk, keeper of the 102- and 106-Series register, reached into his personal archives for pictures of the relevant Berlinas. Thanks also to my old pal and confidante Laurie Caddell of CW Editorial for the 2600 Sprint, GTA and Junior Z pictures. Other shots in the book are by Jon Dooley and Mary Harvey. The 156 *Superturismo* photos were taken by Darren Heath at Imola. Gillian Evans acted as picture editor. Thanks very much to Rod and Judith at Veloce Publishing for having the courage to publish a long-overdue book on my favourite subject. I'd like to dedicate the book to Alfie Tipler.

- John Tipler

4

Alfa Romeo Berlinas

CONTENTS

Thanks 4

Introduction 6
    Committed devotees 9

Chapter 1 The First Berlinas 12
    State occasions 17
    Head-hunted 18
    Factory-built bodies 21

Chapter 2 The Beginnings of Production
Line Methodology.................................... 26
    Exit Jano 30

Chapter 3 Revival of Fortunes 37
    Column shift 37
    An official function 38
    Monocoque construction 42
    Coachbuilt coupés 47

Chapter 4 Pretty Woman 48
    Final evolution 54
    Mini rival 56

Chapter 5 Big Sisters 57
    Straight Six 58
    Shared powertrain 59

Chapter 6 *Bella* Giulia 62
    The T.I. Super 66
    Winning streak 68
    Coupé derivatives 70
    My favourite 71
    The Giulia Super 74
    Grown-up Berlinas 78
    Berlina torque 81

Chapter 7 An Alfa for the People.............. 83
    Shooting Brake 86
    Waiting for Junior 87

Chapter 8 More Metal for the Money......... 91
    Trieste launch 92
    Giugiaro design 93
    The robots arrive 93
    Bumper crop 95
    Nice name, shame about the shape 95
    Range topper 97
    The Barge 98
    The right angle 102
    Your number's up 103
    Extreme evolution 106
    One spark good, two sparks better 107
    Suspension tweak 108
    Family motoring 109
    Monster fun 110
    Blown over 111

Chapter 9 Southern Output 113
    Arna good 113
    Best seller 114
    Real estates 118
    Top dog 120

Chapter 10 Totally Up-front 122
    A 164 to queue for 127
    Silhouette series 128

Chapter 11 Back on Track 129
    Cast-iron case 132
    Quick rack 132
    Superturismo! 135
    Victorious debut 139
    Rise and fall of the Big Bangers 141

Chapter 12 Small is Beautiful 144
    The 33's successor 147
    Facelift 149

Chapter 13 The Thinking Man's
Berlinetta 151
    Tech spec 153
    All fingers and thumbs 155
    User chooser 156
    Starting out 156
    Q-cars 157
    Oil be damned 158
    Common rail system 159
    Diesel fitter 159
    Parallel imports 162
    Track time 164
    Flagship 165

Chapter 14 Saving the Survivors 170
    Tackling the enemy 171
    Spray something simple 173
    Cave dweller 177
    Holy bodies 178

Appendix
    Alfa Romeo specialists & clubs 181

Index 188

Alfa Romeo Berlinas

INTRODUCTION

General awareness of Alfa Romeo saloons / sedans is not good, and people even ask if the 155 is 'the new Alfa', when in fact they mean the 156.

The Alfas that most stick in people's minds are the Spider and the Alfasud. In Europe everyone's mate or uncle had a Alfasud way back when, and they always 'went like stink'. Chances are, when quizzed, folk will manage to remember a 164 or, more topically if they've been watching the ads recently, a 156. But most people won't have much of a clue about model names and numbers. I get people asking me if my 155 is 'the new Alfa'. In my dreams, maybe ...

The point is that Alfa's reputation has tended to rest on its sporting creations and competition achievements, and this is to overlook a huge portfolio of saloons, which go back to the company's earliest days, and which, since the 1950s, have been its mainstay. In the pre-war days, commercial vehicles and marine and aero engines earned the company's living. In the '30s and '40s, the majority of Alfa chassis were available with factory built saloon bodies, and, in addition, were available with coachbuilt bodies in the same way as the exotic sports cars and coupés. In the postwar era, the company took its first steps into mainstream car production with the 1900 Berlina, while the more rarefied Spider and GT coupé versions maintained the sporting image. As *Alfisti* know, the

Spiders were traditionally designed and assembled by Pininfarina and the Giulietta coupés by Bertone. Some saloons - like the 1750 Berlina and Alfa 90 - were tweaked by Bertone, while the Alfasud was drawn by Giugiaro. Only recently was Pininfarina called in to do the business for the 164 model.

In Italy a saloon / sedan is known as a *Berlina* and saloons / sedans as *Berline* (so we've taken something of a liberty in the title of this book in adding an 's' to make the term plural for English speakers). The Berlina name goes back to around 1660, when Italian diplomat Filippo di Chiesa was serving in Berlin, and had a closed four-wheel carriage built for himself, with four-seats in two rows and a cabin flexibly suspended on its chassis. The Berlina was also known as a *Guida Interna*, which implied a closed driving position. Variations on the Berlina theme are the Limousine, a term used from 1905 to denote a luxury car; the Giardinetta, a 'half timbered' shooting brake first defined by Carrozzeria Viotti in 1946; and the Berlinetta, a term coined in 1928 and meaning a two-door coupé. The Familiare estate car or station wagon was current between 1950 and 1980, exemplified by the Colli-bodied Giulietta Promiscua.

One variation on the Berlina layout was the half-timbered Giardinetta *station wagon, in this case built on a 6C 2500 chassis by* Carrozzeria Viotti *in 1946.*

Alfa evolution

At the dawn of the automobile industry, car chassis were originally carried over from the horsedrawn carriage, and constructed from a mix of steel and wooden elements. But by 1910, manufacturers had largely gone over to channel-section pressed steel ladder chassis with riveted crossmembers. By shaping the side members (longerons) to accommodate the customary leaf springs and axles, the ground clearance required was minimised, and these lowered chassis were known as 'ribassato'. This frame construction formed the underpinnings of the very first A.L.F.A.s, and, in fact, remained in use more or less unchanged right up to the advent of the 6C 2300 in 1934.

Up to this point the most fundamental alteration had been the strengthening of the series VI 6C 1750's longeron channel sections with perforated plates. In fact, the 6C 2300 chassis differed mainly in that the crossmembers were welded rather than riveted to the longerons. More radically, the 6C 2300 B of 1935 went over to fully independent suspension, which meant that the rear spring hangers were redundant.

Alfa Romeo's first unit construction model was effectively stymied by the Second World War. A design by Wilfredo Ricart - the five-seater Berlina prototype - was known as the *Gazella*, and the basic shell existed by 1943. The unit consisted of curved front wing tops, with ovoid cut-outs for the head-

lights, a deep V-shaped aperture for the Alfa shield, side grilles in the front panel, plus floorpan, boot floor and inner rear wheelarches. Sadly for posterity, it was abandoned when production got going after the war, and instead we have to console ourselves with the fabulous lines of the 6C 2500 Sport *Freccia d'Oro* . On this model, the bodywork was welded to the chassis, which not only stiffened up the structure, but demonstrated that the thinking behind it was still in the direction of a unitary body-chassis construction.

The first Alfa Romeo monocoque was, of course, the 1900 series that appeared in 1951. Even here the remnants of separate chassis members were carried over in the box-sections

The first mass-produced Alfa Romeo saloon was the 1900 Berlina of 1951, and it was the company's first true unit-construction model as well.

Below right - Apart from purpose-built racing models such as this Zagato-made Coda Tronca SZ of 1962, which had a spaceframe chassis panelled in hand-beaten aluminium, all Alfa Romeos produced after the 1900 model were of monocoque construction.

in the inner wings, scuttle and floorpan and cross braces that formed the rear bulkhead. The basic shell structure provided the framework on which the wings, doors and other closure panels could be hung, as well as the mounting points for suspension and powertrain. Apart from small runs of the racing models like the spaceframe chassis Giulia TZ coupé, all subsequent Alfa production cars were unit construction chassis. However, as safety issues became a concern, bodies with substructures capable of absorbing impacts and crash damage evolved from the 1960s onwards, developed in crash test facilities at the Balocco track into the deformable structures and passenger safety cells that have become mandatory in the 1990s.

With modern construction methods and environmental legislation came new water-based paint systems, a far cry from the original hand-painted bodies made by the coachbuilders of the 1920s. Then, using oil-based paints that took days to dry, painters often stood barefoot in troughs of water to avoid raising dust that would stick to the pigment. Brushmarks were abraded with whetstones prior to polishing and sealing with varnish. By the 1930s, nitro-cellulose paint was in use, and its volatile solvent brought instant drying properties. Preparation by hand and compressed air spray guns were methods still in use in the Portello paint shops, and not until the new plant at Arese opened in 1963 did dipping tanks to coat the entire bodyshell come into use. By the 1980s, electro-plating or cataphoresis had

become the normal way of treating the body-in-white prior to the painting process.

Going back to the beginning, the first generation of Alfas was the product of design engineer Giuseppe Merosi and Antonio Santoni. Although initially ahead of the game, they were pretty conventional machines, not so dissimilar to the rest of the automobile industry's products. Merosi's earlier designs for Bianchi had a sporting bias, which he brought to the very first A.L.F.A. product, the 24HP model. Between 1910 and 1925 then, all Alfas were based on the same chassis layout, while engines were made with aluminium block and cast iron heads, with sidevalve or overhead valve layout, and a gear- or chain-driven camshaft situated in the crankcase. By 1924, Nicola Romeo was urging Enzo Ferrari and his managers Giorgio Rimini and Ing Edoardo Fucito to secure the services of the talented engineer Vittorio Jano. When Jano arrived at Portello in 1925, work began on the new range of commercially viable cars. These were the 6C 1500 models, which were light in weight, efficient, yet functional. The classic Alfa Romeo all-aluminium, four-cylinder, overhead twin-cam engine was born here, and the concept remained virtually unchanged until 1995 when the block construction switched to high-tensile strength cast-iron, with toothed belts instead of timing chains.

For taxation purposes, early models were designated according to maximum power output in HP, or horsepower, which bears no relation to the

modern measurement of brake horsepower or bhp. The formula was made up of the number of cylinders, swept volume, times the engine stroke - whether four- or two-stroke. In practice it encouraged manufacturers to go for long stroke configurations to keep within a lower tax band. In Italy, the four tax groupings were below 6HP, up to 12HP, up to 16HP and under 24HP, and it wasn't long before manufacturers were deliberately underestimating HP ratings to avail their customers of lower duty. Perceived advantages in performance were then attributed to chassis excellence, but as this downgrading of performance ratings became standard practice throughout the industry, HP ratings were revised to include both the original and more realistic rating. Thus we find models designated 15/20HP, 14/16HP and 20/30HP. It's interesting to note that Alexandre Darracq's original motivation for setting up the Portello plant that spawned A.L.F.A. in 1910 was precisely to avoid taxes in France.

Light-alloy synchromesh gearboxes came into use in about 1935, about the time that rigid rear axles and leaf springs gave way to independent set-ups that included coil springs and

Launched in 1972, the Alfasud broke with company tradition in several ways. It was built in southern Italy, it was front-wheel drive and powered by a flat-four engine, but, more fundamentally, was accessible to a much broader market than hitherto.

If you need four seats and four-door practicality, plus sparkling handling and decent performance at an affordable price, you can do a lot worse than the Alfa 75. It's reliable, too.

Alfa's charisma was examplified by the antics of the Autodelta GTAs seen here on the grid for the European Touring Car race on 31ˢᵗ July 1966. Nanni Galli's GTA flanked pole-sitter John Whitmore's Lotus Cortina, with Jochen Rindt's GTA and Hubert Hahne's BMW on the next row. Behind them were the Belgian Team VDS GTAs, with Jackie Stewart's Lotus Cortina and Roy Pierpoint's Mustang also visible, ahead of the hoards of Fiat Abarths and Mini Coopers.

hydraulic dampers. Alfa's advanced mechanical braking system was also phased out with the hydraulic system introduced on the 6C 2300 B of 1935. The effective three-shoe hydraulic drum brakes gave way to Dunlop disc brakes in 1963, first seen on the Giulia T.I., and replaced by the ATE disc system towards the end of that decade.

Original A.L.F.A. wheels were steel-rims with wooden spokes, directly descended from carriages. The pressed steel Sankey wheel was in use by 1913, and shod with beaded-edge tyres that were held in place by the 'C' section either side of the wheel rim. Early tyres were made of overlaid canvas carcass, and gave way to cord tyres, made with twisted cords bonded into the rubber. The first Alfa Romeo to get these was the R.L. Normale of 1922. Another sophistication was the straight-side

wheel, first fitted on the 6C 1500 Normale of 1927, which used a metal ring fixed on the wheel rim to retain the tyre. After 1928, drop-centre wheels were adopted, which, as far as tyre fitment was concerned, were basically the same configuration as today's wheels. Radial tyres appeared in 1946, and Michelin X metal-belted tyres graced the 1900 and Giulietta models from 1952. With rubber coated belt construction, Pirelli Cinturatos were less durable, but gave better roadholding.

Committed devotees

Nowadays the Berlinas have many committed devotees among enthusiast family motorists who value high performance and fine handling at an affordable price - coupled with four-door practicality and decent carrying capacity. What's more, Alfa saloons often handle better than their sporting siblings because their four-door bodyshells are stiffer and less flexible than the coupé or spider models, which, in any case, have generally been based on the saloon car platform.

9

In the early '60s the Giulia T.I. Super was prominent in European Touring Car racing, and paved the way for the more enduring success of the GTA coupé. Like this latter-day Dutch club racer, pictured in the Assen paddock, the model has a following that's stronger than ever in the late 1990s.

The Tipler stable has included a number of Alfa saloons over the years, including this Giulia 1300 T.I., which spent much of its time commuting across town in London in the mid-80s. It proved - more or less - that classic cars could be practical everyday transport.

Often pitted against the Giulia T.I. Super in its early days was the Lotus Cortina, built at the Lotus factory to Chapman's specifications to trounce the rest of the Touring Car field and put Ford on the road to fame and fortune in European racing. Pictured here, harrying Jochen Rindt's Autodelta GTA in Russell Corner at Snetterton during the 1966 European Touring Car 500kms event, is Sir John Whitmore's Alan Mann-run Lotus Cortina.

To an extent the Berlina has been in a Catch 22 situation; rather neglected in the British and North American markets as being too outlandish, or, on the other side of the coin, as lacking the charisma of the sports models. And Alfa's frequently idiosyncratic designs are forgiven or even exalted as extrovert in a stylish GT model, but are less tolerated in a family car. Devoted Berlina fans, though, are blind to such shortcomings, and hold them in just as high regard as the high-profile sportsters.

Whatever the perception of the automobile market at large, there's no doubt that Alfa Romeo was the trendsetter in the sports saloon segment, consistently producing family cars for the enthusiast - what we think of as touring cars today. And while it's true that in the 1950s and '60s other makes like Borgward, Riley and Volvo also made credible and successful sports saloons, Alfa did it for longer, and made sports cars, too. True enough, the 2002 Tii was up with the class of the field in the early '70s, but this only serves to emphasise that today's much-vaunted Bavarian products are mere *arrivistes* on the scene.

I've been a fan of Alfa saloons since the mid-1960s, when my mate

Roger Palmer had a Giulia Super. Having just graduated from those voluminous scooterists' Parkas, with 'Vespa GS 150' or 'Lambretta TV 175' writ large on the back, our new sartorial garb extended to mechanic's overalls inscribed with the legend 'Autodelta SpA'. We fooled nobody, of course, but it was all part of the fun. The frequent Sunday morning routine was off with the hubcaps and tape over the headlights, and maybe zoom down to Brands or up to Snetterton to cheer Rindt and De Adamich *et al* at the European Touring Cars 500kms marathon.

Just what is it about Alfa saloons that puts them ahead of all the other massmarket dross in the desirability stakes? Is it just the heritage factor and power of the Alfa name, or is it a more complex combination of sporting reputation, vehicle temperament, driveability and mechanical make-up? Looks don't come into it with many of them, unless you're one of the converted, and there have at various times been cars equally as swift or with similar performance - a Riley 1.5 against a Giulietta T.I., the Lotus Cortina versus the Giulia T.I. Super, or a Dolomite

The car that really helped restore Alfa Romeo's reputation as a builder of world-class sports saloons was the 156, announced in 1997.

Sprint versus a 2000 Berlina, for example. There's also a rarity factor, in that Alfa Romeo has always made fewer cars than the major players, so you're unlikely to find your neighbour has the same car as you, even if you run a 33, which was the most prolific Alfa.

For a good decade now, the BMW 3-series has been almost unanimously regarded as the benchmark sports saloon, and there's a certain injustice here. Because if you go back to the 1950s and 1960s, when Munich was still struggling to get off two and three wheels and back onto four, it was Alfa Romeo's Giulietta and Giulia Berlinas that set the standard for sporting family saloons. They were compact, lightweight cars, powered by high-revving twin-cam engines running on a suspension set-up that could provide an exhilarating sports car ride and handling. They weren't cheap - you could get a Jaguar for Alfa money in the mid-1960s - but then, an Alfa was accorded the same prestige as a Jaguar, which meant something different then to what it does now; the Jag's Le Mans and Touring Car successes were then recent history. The Alfa also basked in the kudos of not only being a hot-blooded Italian - *la Dolce Vita* and all that - but of being made by the people who'd built the F1 World Championship-winning cars and were having a crack at the European Touring Car

Championship.

Over the years, I've had a number of Alfa saloons, including a 1750 Berlina, a Giulia T.I., a couple of 75s and a 155 as my personal transport, and they've had the immediate advantage of carrying capacity and four-door accessibility. Their shape has always been understated yet purposeful, not immediately obvious to the speed cop, yet without fail intimidating to a rival hot-shoe: any self-respecting boy-racer knew that Alfas were quick as well as surefooted and would want to try his luck. Apart from that, they depreciated to the extent that, given a few years out of the box, almost any model was affordable on the secondhand market. That's not good news if you bought one new, but it is advantageous in that it makes them astonishingly good value. But why buy an Alfa when a used BMW is within reach? Two reasons. One is the feel of the car on the road, the way it responds and flirts with you, engages rather than distances itself from you. It's an involving experience, rather than a sterilised encounter. Your Alfa saloon may be flawed one way or another, but it has bags of charm and character, which even the above-average family car lacks. Just place a 156 alongside a 3-series and see which one has the dynamism of an animal about to pounce, and which one looks like a close relative of Robo-Cop. And, forgive

a touch of smugness, but there's that stigma thing that the Alfa never suffered from. It's a yuppie hangover, if you like, which puts the uptight materialist behind the wheel of a Scando-Teuton, something you could rarely accuse an Alfa owner of being. The BMW and Audi A4 are the safe bet for those of a sober disposition, while any Laguna, Prelude, or Primera not prepared to BTCC spec is, in my filing system at any rate, pigeonholed under repmobile along with Mondeo and Vectra, V6 engine or not.

When the 156 was launched in 1997 to rave reviews, industry pundits were confidently predicting that it would dominate the sports saloon market, ousting BMW and Audi from the top of the tree. It hasn't quite carried that off, but it did succeed in thawing many hearts of ice, and still the motoring media keeps willing Alfa to succeed. In July 1999, Channel 4's excellent *Driven* programme featured a three-way test between a Lexus IS 2000, a 3-series BMW and an Alfa 156, and the Alfa comprehensively outshone these particular rivals. And as I write, the 166 is regularly ranged against cars like the BMW 528 and Jaguar S-type in executive shoot-outs. Whilst the BMW seems to come out on top on the basis of residual values and build quality, the Alfa is invariably the outsider's choice; in fact, it's the wild card in the pack. I like that, and it would almost be a shame if Alfa was to become top dog.

- **John Tipler**

1

THE FIRST BERLINAS

Alfa Romeo's history as a manufacturer of saloon cars goes right back to the company's origins, so that's where we'll begin the story. One or two definitions are in order here.

Sometimes you'll find a model referred to as a *Berlina*, and that's simply because, in Italy, the term Berlina means a closed saloon car, as does the expression *Guida Interna,* which was used in the 1920s and early '30s to describe a car with fully enclosed bodywork. The first cars produced by A.L.F.A. in 1910 - the Merosi-designed 24HP and the more economical 12HP models - were available with *Limousine* coachwork, so that makes them saloons of a kind, although they were more regularly clad in open top 'torpedo' or semi-closed *Landaulet* body-

Here's a contemporary illustration of what the Portello factory looked like in 1910, its roof emblazoned with the A.L.F.A. acronym (Anonima Lombarda Fabbrica Automobili). A contemporary 24HP Phaeton stands outside the building.

Exterior of the foundry block at Portello, circa 1916, with water pipes ready for installation, plus a batch of chassis to the right.

work by local *Carrozzerie* such as Castagna: a deliberate ploy to acquire an upmarket image.

The 24HP model was a relatively large and imposing car, powered by Merosi's 4.0-litre, four-cylinder monobloc engine linked to cardan shaft transmission, which was in production until 1914 in four series (A, B, C, D) with minor modifications. In that time, 301 24HP chassis were made. Don't forget that we're dealing with very small production volumes, decades in advance of mass-production. The less complex 12HP was made until 1911 (49 units in all), and, in the company's first foray into competitive motoring, a 12HP car tweaked to 25HP tied with five other competitors for first place in the 1500km Modena trial in April that year. The torpedo-bodied car was driven by test driver Nino Franchini. Allied with an upmarket image, it was perceived that competition success would swiftly consolidate the company's reputation.

However, a short wheelbase version was available in *Fiacre* or taxi format, designated the 14/16HP, from 1910 to 1911. The 12HP soon evolved into the longer wheelbase 15HP of 1912 (99 units built), with a four-speed gear-box replacing the three-speed unit, and mounted separately from the engine - as was customary - underneath the front seats.

By 1913, A.L.F.A. customers were demanding higher performance than the 24HP could supply, and Merosi was asked to produce a more powerful car. The 40-60HP model was the result, and it was powered by a sophisti-

The performance of Count Marco Ricotti's 40-60HP Aerodinamica of 1914 was considerably better than the standard model - 30kph, in fact - and it's likely to have been even better if the tear-drop shape had been the other way round.

A.L.F.A. 15HP engines being bench-tested at Portello around 1913. The intake and exhaust manifolds are prominent, as is the distinctive angled radiator connection on top of the engine.

cated 6.0-litre overhead valve pushrod engine with two camshafts housed in the crankcase. A substantial shroud that hung below the engine directed cooling air generated by the fan. Of the 28 units built, a number were equipped with *deux baquets* bodies - in other words, two-seater roadsters - for competition use. One, however, was fitted with the most outrageous closed bodywork which looked like a giant teardrop, and was surely the boldest attempt to create an aerodynamic body on an Alfa platform until the B.A.T. cars of the 1950s. Built by Carrozzeria Castagna to a design by Count Marco Ricotti, the futuristic egg-shape body was mounted on a 40-60HP rolling chassis. With porthole windows and

wraparound windscreen, this proto-spaceship looked as if it was a forward-control vehicle, but the driving position was still located in the usual place, midway down the chassis. Whether its aerodynamics worked that well is debatable, as it might have been better with the pointed end in front. The maximum speed of 140kph was actually some 30kph better than that of the standard model, however. This, then, was the first Berlina, albeit a pretty bizarre one. Known as the 40-60 Aerodinamica, it is a key exhibit at the Alfa Romeo museum at Arese, and the audacity of its conception is really quite breathtaking.

During the First World War, the company came under control of Nicola

Romeo, who owned a flourishing agricultural tractor, traction engine and marine engine business. Romeo was now A.L.F.A.'s chief shareholder, and produced munitions, aero engines (under license from Isotta Fraschini), and compressors derived from the 20-30HP engine. After the war, A.L.F.A. production got going again with the 15-20HP model that had appeared briefly in 1914. The 20-30HP, which was used in active service as an ambulance, was effectively the revised 24HP. This, too, had appeared first in 1914 and resurfaced in 1920 using 24HP components.

During 1920 there were turbulent times within the company as working practices, unionisation and finances were sorted out, but evidence of corpo-

The Portello plant was light and spacious, complete with aerial gantries and gas lights, as seen in this shot of the steel rod stores in about 1912.

rate unity was manifest in the 20-30HP, the first car to be badged as an 'Alfa Romeo'. There was also a sports model designated the ES Sport, and a racing version was equipped with twin-spark ignition. The latter achieved success in the hands of Baroness Maria

The handsome 20-30HP of 1921-22 was essentially a pre-war design revitalised as production struggled to make headway against a background of financial and industrial confusion within Nicola Romeo's corporation. Note the continued use of artillery-type spoke wheels.

The 6.3-litre six-cylinder 30-50HP G1 limousine was made between 1921 and 1922, and was typical of the work of Giuseppe Merosi, although the inspiration for the engine had come from Enzo Ferrari. Just 52 examples were made.

Antonietta Avanzo, who finished third in the Brescia Grand Prix of 1921 with Giulio Ramponi acting as riding mechanic. But, for our purposes, the only saloon body on offer was a Limousine.

The Merosi-designed, 6.3-litre, six-cylinder G1 - also referred to as the 35-50HP - was available in 1920 as a Limousine and, of totally different configuration, as a Spider - (the term, incidentally, coined by Dublin coachbuilder Archibald Holmes for a spider-like, open-wheeled, two-seater carriage back in 1860). As a means of reducing unsprung weight, the chassis was equipped with a double set of cantilevered leaf springs, mounted on the rear axle at one end and halfway along the chassis at the other. A longitudinal thrust rod ran parallel to the propshaft to aid axle location. More significantly, encouragement for this powerful engine had come from Alfa Romeo works driver Enzo Ferrari, who visualised competition success with it. Rather oddly, though, all 52 units built were exported to Australia, partly because it was not an economical car to run in Italy. A limousine version, clearly aimed at the top end of the market, was shown in London in 1921.

Further constraints on production during 1922 resulted from the collapse of the Banca Italiana di Sconto - to which the company was in hock to the tune of 90 million lira. The intervention of Mussolini's fledgling government, through the Banca Nazionale di Credito the following year, put matters to rights, and the company was effectively supported by state aid. At the same time, Alfa's reputation was bolstered by victory in the 1923 *Targa Florio*, which prompted adoption of the four-leaf clover *quadrifoglio* symbol that adorns top-of-the-range models today. The Targa Florio win was also the incentive for Merosi to build the 2.0-litre Grand Prix car that came to be known as the P1, driven by Sivocci/Marinoni. This capacity reflected the switch from a 3.0-litre formula, which - paradoxically - had the effect of scuppering the R.L. model's chances as the basis for a Grand Prix car. The R.L. was available in Limousine and torpedo form, designated the *Normale*, and the rather more dashing Sport and Super Sport versions could also be bought with these body styles as well as the two-seater Spider, in which guise it enjoyed considerable success in road racing. You can see some of the sports and cabriolet versions of the R.L. at the Alfa museum.

State occasions

By 1925, Mussolini was sufficiently impressed by Alfa Romeo's sporting accomplishments to make the cars his official state transport. There were several variations on the R.L. theme, including an interesting four-door 'Faux Cabriolet' by the English Carlton Carriage Company, which featured fake rear hood irons that were purely ornamental. Mostly, though, the cars were clad with coachwork by Castagna, Alessio, Falco, Sala and Zagato. Chassis exported elsewhere inevitably ended up with locally produced coachwork, and the five body styles available in the UK were possibly rather more restrained than the native Italian versions. An R.L. Super Sport with a four-door saloon body cost £1025 from the British concessionaire, established at Portman Square, London W1, in 1925, making it the most expensive of the styles available. The idea that you pay more for a sports/GT model had, evidently, yet to catch on.

The R.L.'s overhead valve, 2916cc, six-cylinder engine was ahead of the game, while other innovations it ushered in were wire-spoke wheels that replaced the pressed steel Sankey variety, plus adoption of front brakes in 1923. The model evolved through five consecutive series, with minor mechanical uprating, to 1925. The final evolutions also included the Super Sport version, distinguished by an imposing V-shaped radiator grille that had the Alfa badge at the top on both facets of the V, as well as the Alfa Romeo script low down on the righthand side. The R.L. Super Sport was fitted with flat-topped aluminium pistons and modified engine casings, and breathed through twin Zenith 42 DEF carbs fed by a camshaft-driven air pump that maintained pressure in the fuel tank. Altogether, 1849 units of the R.L. Normale and Sport versions were produced, along with 521 units of the

Left: Look at the radiators of these 20-30 ES Sports racing cars - and the P1 in the middle - and in 1923 they were still badged as Alfas, although Nicola Romeo had owned the company since 1915. Meanwhile, Enzo Ferrari, at the wheel of the car second from left, came second in the Circuit of Mugello in 1922.

The shot of this R.L. Super Sport of 1925 shows something of the quality and grandeur of the cars that Alfa Romeo was producing at the time. The coachwork in this case is by Castagna, while others were done by Zagato and, if the model was exported, very likely by a domestic coachbuilder.

Below, left: The dashboard of the R.L. Super Sport of 1925-27 contained Jaeger instruments, including, from left to right, the rev-counter, odometer, oil pressure gauge, ignition and headlight switches, the fuel pressure and level gauge, plus a clock. The gear lever and the hand-brake lever are at centre, while the ignition advance and hand-throttle levers are on the steering column. Clutch and brake pedals are embossed with the A.L.F.A. logo.

cylinder car with aluminium crankcase and cast iron cylinder block, which used much of the R.L.'s componentry but was a simpler - and therefore cheaper - model. The longer chassis of the Unificato was best suited to carrying saloon-type bodies. There's an R.M. Sport spider in the collection at Arese.

Head-hunted

Apart from Merosi, and, of course, Nicola Romeo, the most influential figure at Alfa Romeo at this time was Enzo Ferrari, who, in the finest traditions of behind-the-scenes corporate poaching, head-hunted a couple of important engineers from Fiat. One was Luigi Bazzi, who remained with Enzo until the late 1960s, and the other was Vittorio Jano, who had masterminded Fiat's postwar racing cars, including the 805/405 of 1922. When Jano came out with the Alfa Romeo P2 in 1924, his former employers were sufficiently incensed to have his premises searched to see if he had taken the Fiat drawings with him. The P2's success was largely responsible for the sidelining and resignation of Merosi, and the elevation of Jano.

R.L. Super Sport.

By 1926 the R.L. Normale was relaunched as the R.L. *Turismo*, fitted with detuned Super Sport powertrain and much of that model's ancillaries, including bigger brakes and now featuring nickel-plated brightwork. The saloon versions were inevitably limousines or *Coupé de Villes*, in which the chauffeur in the front seat was not under cover. 600 units of the R.L. Turismo were made.

A far more straightforward machine was the R.M., produced as Normale and Sport models, plus the longer wheelbase *Unificato* between 1923 and 1926 - with just 278 cars built. The R.M. was a 2.0-litre, four-

A contemporary poster advertising the six-cylinder 2994cc R.L. Turismo of 1926-27: Art Deco styling very evident. This model was also available as a limousine.

It's clearly not a saloon, but this four-cylinder, 1996cc R.M. Sport of 1924 with a short-chassis spider body was available with closed Guida Interna bodywork.

Torpedo Sport (con capotta incassata) su chassis R. M. Sport

Another variation on the cabriolet body style was known as the Torpedo, in which the soft-top of this R.M. Sport was encased within the rear bodywork.

Jano's first production car was the 6C 1500 Normale, launched in October 1926 at the Paris Show. Low chassis weight, combined with a high performance engine, reflected the P2's core values, whilst, at the same time, the model was more widely affordable than were the R.L. and R.M. The 6C 1500's pressed steel chassis was the same length as that of the R.M. in four-seater format (2900mm), but longer by 200mm (3100mm) for the six-seater body. The *guida interna* saloons were most frequently fitted with sedate, almost austere Weymann coachwork, which sometimes featured fake cabriolet hood irons in the rear three-quarter roof pillar. The trunk, or luggage boot, was tacked on almost as an afterthought. In terms of its fabrication, the Weymann body featured a leathercloth cladding over a wooden frame: a traditional system patented in France. It had the advantages of light weight and flexibility, as well as being less resonant than a steel panelled construction. An extension of this format was the semi-rigid body, in which only the upper part of the body was constructed using the Weymann method with the lower half, from the waist down as it were, made in steel.

The 6C 1500's ladder frame chassis curved elegantly over the rear axle and was also curved downwards at the front. Suspension was by semi-elliptic leaf springs front and rear (with friction dampers) and a torque tube at the back axle. Drum brakes were fitted all round.

The tall 1487cc, six-cylinder unit was fed by a single vertical Zenith 30 D2 carburettor, and consisted of a cast iron block and cylinder head, aluminium sump and crankcase, with an overhead camshaft and tappets adjusted by a couple of discs that were screwed on to the valve stem. This was a system derived from Hispano Suiza and a feature of the majority of Alfa Romeo engines until the Giulietta Sprint appeared in 1954 with bucket-type tappets and spacers. The 6C 1500 was one of the first production cars to be equipped with an overhead cam engine, which service outlets had some difficulty setting up at first. The 12-volt electrics included an ignition system incorporating a distributor and battery, Bosch starter motor and dynamo, and cooling was augmented by an aluminium fan. Straight-side wire

spoke wheels were fitted with either 28 x 4.95 or 30 x 5.25 Superflex Cord beaded tyres.

For 1928, Alfa Romeo launched its 1487cc, overhead twin-cam unit in the 6C 1500 Sport model. This landmark car was not only Alfa's first production twin-cam, but also an engine configuration virtually unheard of outside the realms of competition motoring. It developed 54bhp at 4500rpm, compared with the 44bhp at 4400rpm of the single cam unit. In the form of the supercharged Mille Miglia Speciale spider that appeared in 1928, it developed 76bhp at 4800rpm. The latter used a Roots-type blower made at Portello, allied to the specially tuned Zenith 30 D2 carburettor, and in order to house the unit the engine was set back in the chassis. The shorter wheelbase Sport model chassis was basically similar to that of the Normale, although the length of the leaf springs differed. While the Berlina versions still appeared quite restrained, the spider bodies were becoming much more flamboyant, as demonstrated by the Zagato coachwork, and verging on the grandiose in some cases, such as the offerings by Touring and Garavini.

20

This is a 6C 1500 with a semi-rigid Weymann guida interna *body by Nord Italia of Milan. Items of note are the false cabriolet irons on the C-pillars, round door handles, split-opening windscreen and fairly crudely made front bumper. It's clearly a more utilitarian vehicle than the spider.*

Below, left: The first significant model to be designed by Vittorio Jano was the 6C 1500 Sport, which came out in 1927. It combined all the virtues that we look for in a sports saloon, namely light weight, manoeuvrability, low maintenance and high performance. This is the spider version - it was also available with a torpedo or guida interna *body.*

Factory-built bodies

The International Motor Show was held in 1929 in Rome, and was chosen as the obvious event at which to launch the 6C 1750 Turismo, a revised and uprated version of the 6C 1500 single overhead cam unit. Both bore and stroke were increased to bring capacity up to 1752cc, an approximate displacement that Alfa seems almost routinely to have returned to with successive model ranges. A Solex 26-30 FFV vertical carburettor was now used, and power output was 46bhp at 4000rpm. The 6C 1750 Sport model was fitted with a twin-cam version of the same engine, developing 52bhp at 4400rpm. The valve-gear was driven by a vertical shaft and twin pairs of bevel gears, with the lower one keyed onto the five-bearing crankshaft. The transmission casing was cast in aluminium, and hemispherical combustion chambers permitted larger valve diameters as well as a central location for the spark plugs.

This new 6C 1750 Turismo was available in open-top Spider and Torpedo format, as well as a variety of closed Berlina body styles, including Weymann and Dorsay Coupé de Ville limousines. From this point, there was a concerted effort to build the Berlina bodies in-house at Portello, and, perhaps predictably, the factory bodies

A total of 1049 units of the 6C 1500 Normale and Sport were made.

But if 1928 was a memorable year for Alfa's model range, it was also a time of major personnel changes. Nicola Romeo resigned as managing director and left the company, as did his close aids Edoardo Fucito and Giorgio Rimini, and Ing Pasquale Gallo, who was succeeded as general manager by Ing Prospero Gianferrari. Meanwhile, Vittorio Jano carried blithely on, basking in the success of Campari and Ramponi, who won the 1928 *Mille Miglia* in the eponymous supercharged 6C 1500 spider.

The following year, Scuderia Ferrari was established at Modena to run Alfa Romeo's competition department, whilst at a more prosaic level, the T500 Alfa Romeo *Autobus* was available on the commercial vehicle market. This was a long-bonneted coach characterised by a roof ladder and three windscreen panels. And as a lucrative sideline, the company made Armstrong Siddeley aero engines under licence for Caproni, as well as a range of marine engines of up to 500HP. Now, my wife has an ex-naval harbour launch - an HLD - and it's a fine little ship, although I can't help but reflect how much more agreeable it would be if its Foden engine was replaced by an Alfa Romeo unit!

Inset: The 6C 1750 model of 1929-33 was available with Berlina, spider and torpedo body styles, and this is a rather more exclusive six-seater Dorsay Limousine version, in which the driver is uncovered and separated from those privileged to travel in the rear of the car.

Eusebio Garavini founded the Carrozzeria that bore his name in Turin in 1914, and gained a reputation for the excellence of his Weymann semi-rigid wood-and-canvas bodies, exemplified by this 1929 6C 1750 Turismo. However, the firm didn't recover from the devastation of its factory in the Second World War.

were plainer than those made by Felice Bianchi Anderloni's Carrozzeria Touring Superleggera concern, whose *Mignon* model took the gold medal at the 1930 Rimini *Concours d'Elegance*. Touring's renowned 'Superlight' construction consisted of a slender steel skeleton mounted on the chassis, which was clad with lightweight aluminium panels. What was considered the norm for sports and competition cars was equally appropriate for Berlina construction, if rather over-the-top when applied to a grand ceremonial vehicle.

It did help with performance, however, since there's nothing like a lightweight body to provide an instant gain in performance.

For chassis destined to be bodied outside the factory, Alfa Romeo supplied a set of 1:5 scale drawings giving

6C. GranTurismo

This 6C 1750 Berlina has a factory-made body, constructed by the traditional Weymann method. There was no set pattern for location of the door hinges, and here, the front door is hinged on the A-post, and the rear door is hinged at the B-pillar. By 1929, the merits of coachwork produced in the Portello bodyshop were being actively promoted.

all the relevant dimensions for the coachbuilders' reference. Some 6C 1750 Turismo models featured luggage boots, which had a pair of spare wheels hung on the back, whilst others carried their spares mounted flush with the upright rear end of the car. Some of these cars had front hinged doors, some were hung on the central pillar. Hinging the front doors at the front on the A-post and the rear ones at the back on the C-pillar, so that the handles met in the middle allowing easier access to the interior. Another

The opulent interior of a 6C 1750 guida interna three-light saloon of 1929 to 1933. The driver is separated from the occupants of the rear by a glass partition.

There would have been two jump seats, which folded away to allow easier access and greater space.

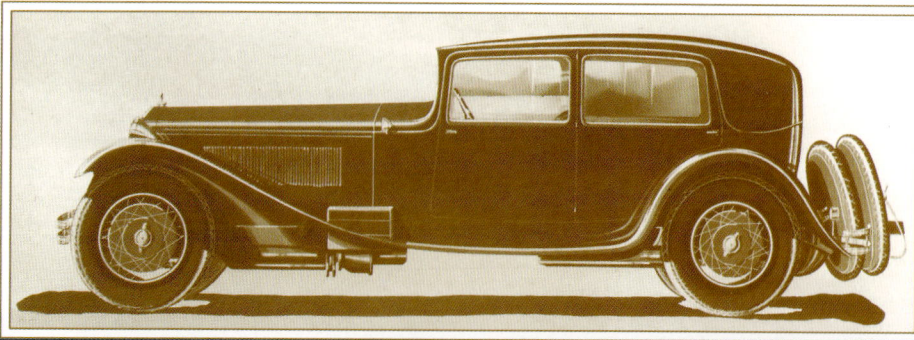

The gloomy and somewhat rough and ready Portello foundry department, pictured during the 1930s, when a variety of castings were produced.

more spacious variant, known as the three-light saloon, contained a glass partition that separated the driver or chauffeur in the front from the passengers behind, and featured a pair of collapsible jump seats in the rear, rather like a modern taxi. Turin-based Carrozzeria Eusebio Garavini was the noted exponent of this style.

The 6C 1750 Turismo's drum brakes were actuated by rods, and the central accelerator pedal divided the clutch and brake pedals. It's interesting to note that all Alfa Romeos were right-hand drive, right up to the 1900 that came out in 1950. At 3100mm, the 6C 1750 Turismo's wheelbase was the same length as that of the 6C 1500 Normale, and the chassis' raw material was supplied by Cogne. Brightwork was nickel-plated and metal ancillaries were chromed. The 6C 1750 Turismo ran on straight-side wire-spoke wheels, shod with Pirelli Superflex Stella Bianca tyres. Externally, there wasn't much that distinguished the 6C 1750 Turismo from the 6C 1500 in the more prosaic body styles, although the larger capacity model had narrow polished rims to its Bosch headlight surrounds. Items like the dashboard were made in-house, and the white-faced Jaeger instruments were produced to Alfa Romeo designs.

24

This Portello-bodied 6C 1900 Gran Turismo Berlina of 1933 ushered in the new six-cylinder engine that was now bored out to 1917cc with an aluminium cylinder head. Further innovations included a box-section chassis with apertures on the inner face of the longerons, plus wider track front and rear and beefed-up rear springs.

In total, 1131 units of the 6C 1750 Turismo were made, with 277 of the 6C 1750 Sport version.

To further complicate matters, Alfa released the 6C 1500 Super Sport in 1929, which was basically the 6C 1750 in its third series incarnation with reduced bore and stroke; the same, in fact, as the 6C 1500. Confusingly, a few were increased to 1752cc and fitted with fixed cylinder heads and seven main bearing engines. The one-piece block and head unit was robust enough to cope with the stresses imposed by a supercharger - there was no head gasket to destroy. The main purpose of this model was to contest the 1500cc class, but it was also available with the Berlina body if required.

The slightly more powerful 6C 1750 Gran Turismo was introduced in 1930, and a full range of body styles could be specified, with Weymann-type Berlinas produced by Touring as well as at Portello. Factory-built bodies carried a small badge of identification on the lower section of the car's front side panels. The factory model's cabin superstructure was rather taller and less elegant than that of the specialist coachbuilder, and as an early example of what we generally perceive as a GT car, Touring also made a very attractive two-door *Berlinetta* version, constructed by the semi-rigid Weymann method. From 1930 to 1932, 723 units of the 6C 1750 Gran Turismo were made, and there's a two-door model at the Alfa museum.

Performance of the 6C 1750 was lifted in 1931 with the *Gran Turismo Compressore* model, which satisfied owners who needed a big car as well as more power. These supercharged models were built on longer, wider chassis, and to emphasise the sporting potential, a Touring-bodied Berlina G.T.C. took a class win in the 1932 Mille Miglia in the hands of Nando Minoia. Stylistically, the Berlina bodies, particularly the Portello made ones, were becoming rounder at the edges, while the chief mechanical differences were adoption of an in-unit housing for the camshaft drive within the cylinder block, twin choke Memini DOA carburettor and a larger clutch. The friction dampers could be adjusted by way of two enormous knobs that controlled a S.I.A.T.A. hydraulic system located below the steering column, and the track width was broadened by the simple expedient of locating the spring hangers outside the chassis longerons.

For 1933, the six-cylinder engine was given an aluminium cylinder head, and bore was increased from 65mm to 68mm to give a capacity of 1917cc. What turned out to be an interim model was designated the 6C 1900 Gran Turismo (199 units produced), and in Berlina form it could be specified with a factory body painted in an attractive two-tone colour scheme. One of these, finished in two-tone cream and buff colours, is the earliest Berlina on display at Arese, apart from the one-off egg-shaped 40-60 Aerodinamica.

Meanwhile, another model line had been introduced, with a short wheelbase chassis for competition and longer chassis for Berlinas and tourers. This was the 8C 2300, powered by Jano's outstanding straight eight twin-cam engine. Using an aluminium alloy block with steel cylinder liners, the crankshaft was split in two sections by the central drive gear to the cams and supercharger, and the engine was fed by a Memini SI 36 gravity feed carburettor. Specification was gradually improved until 1934, by which time styling had become more elegant, with a slightly protruding V-shape to the lower part of the radiator grille. In all, 249 units of the 8C 2300 were made. Meanwhile, behind the scenes, some major changes were afoot ...

25

Alfa Romeo Berlinas

2

THE BEGINNINGS OF PRODUCTION-LINE METHODOLOGY

Launched at the Milan Motor Show in April 1934, the 6C 2300 was altogether more modern-looking than the gawky upright cars of the previous generation. The fireguard shape of the radiator grille, in this case with fan-like slats, was also symbolic of the decade. This is the 1935 Touring-bodied model known affectionately as Soffio di Satana, *meaning Breath of Satan.*

Times were still somewhat turbulent at Portello when production line methodology began to be ushered in, and it was largely the adage of the 'new broom' that instituted new procedures, along with a 'no pain, no gain' philosophy. The company was taken out of the hands of the Banco Nazionale di Sconto in 1933, and passed on to the Institute for Industrial Reconstruction - the I.R.I. - and its new general manager was named as Ing Ugo Gobbato, an internationally renowned industrial management specialist.

Gobbato's initial strategy meant imposing swingeing job cuts that reduced the workforce from 2100 to below a thousand. However, his radical modernisation programme brought Portello out of the realms of handcrafted products and into the modern world of production-line methodology, manifest in models such as the 6C 2300 B Gran Turismo that came out in 1935. The downscaling of the workforce didn't last long, and by 1940 some 14,000 workers were employed, which is a measure of Ing Gobbato's success. As

a comparison, forty years on, the workforce totalled 45,825 employees in 1980.

At this stage, the Portello plant consisted broadly of six main blocks, with a host of orderly triangular rooflines resembling corrugated cardboard, housing the various production departments - foundry, machine shop, engine and mechanical construction, chassis shop, bodywork, and assembly, as well as administration buildings. Contemporary illustrations show that the factory was mostly surrounded by countryside.

First fruit of the new regime was a new model, launched in 1934 and designated the 6C 2300 Turismo, powered by the 2309cc 68bhp straight six unit. Although its chassis construction was merely a development of the 6C 1900, it represented a significant step forward in that the gear-driven valve mechanism was replaced by chain drive. Attention to detail was of the same standard bestowed on the 6C 1750, with chassis brightwork nickel-plated and certain items of coachwork

Rear three-quarter view of the Touring-bodied 6C 2300 'Breath of Satan', showing off its distinctive beetle back and enclosed spare wheel recess.

Below: The 6C 2300 Gran Turismo came out in 1934 and was available with the whole range of body styles, including Berlina, Cabriolet, Coupé and Spider. The Berlina form shown here was built in-house and upholstered in fabric like the regular Turismo version.

Below: The sporting Gran Turismo model shared the nickel-plated brightwork with the basic 6C 2300 Turismo, including the characteristic front bumper, which featured an elegant linear pattern and a dip in the centre to accommodate the starting handle.

By 1935, Pinin Farina was well on his way as a constructor and car stylist, and this is his commission for Theo Rossi di Montelera, based on a design by Mario Revelli di Beaumont. It's built on one of the 535 6C 2300 Pescara chassis, and was shown at the Milan Show that year. Pinin Farina's client also owned a slightly more conventional-looking Pescara Berlinetta with a superb lozenge-shaped tail, of which Touring built 60 examples to commemorate Alfa Romeo's 1-2-3 triumph in the Pescara 24-Hours of 1934.

and interior door handles chromed. Drop-centre wheels were fitted, shod with 6.00 x 19 tyres. One of the first renditions of the beetle-back look that came to be widely espoused during the following two decades by everyone from Volkswagen and Tatra to Vanguard and Jowett, was the 6C 2300 Turismo semi-rigid, three-light saloon by Touring, done in 1934. This design was nicknamed 'Satan's Breath', and actually came close to being productionised by Alfa Romeo. More typical and rather more conventional was the handsome Berlina of the same date styled by Pinin Farina, which had more flowing wings and mudguards. Altogether, 222 units of this model were produced in various formats.

Rather more plentiful, with 535 units built, was the short wheelbase 6C 2300 Gran Turismo. This version was available in all body styles, with Berlinas clad by the factory with pillarless all-steel two-tone coachwork that was more rotund than hitherto,

and even described as aerodynamic, which it probably was for the time. The front and rear mudguards were fuller than previously, and were starting to look more like wing panels rather than just mudguards. The ostentatious front bumper was engraved with its own coachline, and dipped at the centre to give access for the starting handle. Twin windscreen wipers were fitted, as were semaphore arm side indicators, and the bonnet sides had six flaps for cooling. To be sure of seeing one in the metal, as it were, there's a black one on display in the Alfa museum.

The company was heavily involved in motor racing, of course, and, on the whole, it's that legacy that continues to ensure Alfa road cars are such fun. Victories in major events like Le Mans, the Mille Miglia and the Targa Florio, came thick and fast, mostly involving Nuvolari, Sommer, Brivio and Varzi in the P2, 6C 1750 GS and then the 8C 2300 and P3. Although, in the notoriously high attrition rate of the time,

they lost Campari and Borzacchini at Monza in 1933. A competition derivative of the 6C 2300 Gran Turismo came in the shape of the Touring-bodied Pescara Berlinetta, which featured wonderfully rounded front and rear wings and a lozenge-shaped tail. This was a limited run of 60 cars built to replicate those driven by a three-car team that scored a 1-2-3 victory in the 1934 Pescara 24-Hours. One characteristic was the narrow chromed twin bars that ran vertically down the radiator grille. Pinin Farina came up with a novel design in 1935, cladding one of these Pescara chassis in curvaceous competition-like coachwork, with hardly a straight line to be seen. Headlights were set low in the panelling between the wings, and there was what can only be described as a massive searchlight behind the heavily slatted radiator grille. The aerodynamic appearance was completed by conical wheel discs.

While sports cars like these basked

The 6C 2300 B was equipped with independent suspension and a pair of single choke side-draught carbs (instead of a single twin-choke downdraught version). This four-door saloon was the Portello-made all-steel Berlina. Just 105 units were produced between 1935 and 1937, including a number of stylish Berlinettas and cabriolets.

in the limelight, it's always worth bearing in mind that Alfa Romeo production included commercial vehicles, and contemporary with racy creations like the Pescara model were heavy-duty trucks like the stolid Tipo 50 and Autobus Tipo 85. Promotional copy described the Alfa Romeo Tipo 500 truck as 'A true friend', and adverts featured a large, slavering dog superimposed on the long-nosed truck. An Alfa Tipo 85G truck won the Rome-Brussels-Paris International Competition for trucks running on 'producer' gas (the product of stoves mounted behind the cab and fired with charcoal burners), staged in 1935 in the quest for a cheap substitute for petrol. Towards the end of the decade, this method of combustion - *gassogeno* - was quite common, and some of the big three-axle Tipo 110 Alfa Romeo passenger coaches were so converted.

Despite what now seem rather archaic methods of fuel procurement, things were moving forward meanwhile on the styling front. The fulsome rounded wings that characterised the Pescara Berlinetta were transposed onto the 6C 2300 B Gran Turismo, produced between 1935 and 1938. On the Berlina, the headlights were al-

The 6C 2300 B Turismo was only made as a six-seater Berlina, with just 81 units built, mostly in 1937. Built with steel panels on a C-section pressed steel ladder chassis, this was a large, prestigious car. The spare wheel was housed within the back panel, and could be covered with a metal wheel disc.

most - but not quite - integrated into the wings, and the radiator grille was now more rounded and reclined backwards, while the roofline and rear of the car were still more curved off than the previous 6C 2300 Gran Turismo. More significantly, though, the 6C 2300 B was a complete redesign, apart from its engine, and it had a new chassis with independent suspension all round, with the rear axle flexibly mounted on the chassis. Hydraulic brakes were a further innovation, replacing the mechanical system used previously.

While the Dubonnet-type front suspension set-up on the 2300 B was to Jano's own design, the swinging-axle rear-end was inspired by Dr Ferdinand Porsche. In practice the rear suspension proved unreliable and caused the tyres' angle of contact with the road surface to change constantly, as well as inducing a rear-wheel steer effect. The reason for this was the 45-

degree axis on which the wheels oscillated; consequently, roadholding was never as good as it should have been. Competition cars overcame the problem to an extent with stiffened suspension and by limiting the amount of travel. In fact, the situation was only satisfactorily addressed when Alfa reverted to a live axle with the postwar 1900. In four years, only 216 units of the 6C 2300 B Lungo were produced, which really fails to indicate any quantum leap in production at Portello, and also suggests that Alfa's principal clientele was still the well-to-do enthusiast who required a state-of-the-art sporting car.

A further 105 Pescara derivatives were made between 1935 and 1937, in Berlina as well as Berlinetta form. Like the earlier version, the Pescara used two single choke horizontal carburettors as opposed to the single twin choke carb fitted on the regular 6C

Mechanical uprating of the second series 6C 2300 B included a new gearbox with synchromesh on 3rd and 4th gears, as well as a mechanical fuel pump instead of the original electric one. Pictured is the Corto, or short-chassis version, of which 198 were made between 1938 and 1939. Note the wipers are still housed above the windscreen.

2300 B. A number of Pescara chassis were bodied by Pinin Farina and featured extraordinarily exotic styling in the then-current *Grand Routiére* fashion, while a 1937 offering by Ercole Castagna bore a certain resemblance to the Cord 810 and Auburn 852 that had emerged from Gordon Buehrig's work for Duesenberg a couple of years earlier.

While the 6C 2300 B Gran Turismo was a compact car, the 6C 2300 B Turismo was made only as a factory-bodied Berlina, between 1936 and '37 (81 units in total). This car was a much larger proposition, with seating for six and altogether different proportions, thanks to a sizeable rear three-quarter window. Its spare wheel was stowed neatly within the flat curve of the car's rear panel, and was covered by an appropriate disc that enveloped wheel and tyre.

Posed in front of a suitably portentous military backdrop is this four-seater 6C 2300 Berlinetta, a one-off design by Pinin Farina from 1937, featuring circular Art Deco motifs along the sides of the bonnet.

Exit Jano

With the appointment in 1937 of Wilfredo Ricart as technical consultant, Jano left the company under a cloud. Despite the remarkable success in the previous year's Mille Miglia when the 8C 2900s scooped the first three places, his 4.5-litre twin supercharged 12C Tipo 37 was not a success. Even Nuvolari, driving a more dependable 12C Tipo C36, managed only seventh place at Livorno in the Italian Grand Prix against the German Silver Arrows. However, one important effect of Jano's presence that continued to be felt after his departure was the expansion of his production programme, and Portello continued to manufacture the second series 6C 2300 B, now specified as *Lungo* and *Corto* - long and short -

30

Carrozzeria Touring -Milano-

Berlinetta Touring 1000 miglia

Peso della carrozzeria Kg.126.

Velocità Km.170.

Pictured in an official promotion by Felice Bianchi Anderloni's Carrozzeria Touring, the wonderfully elegant 6C 2300 Mille Miglia Berlinetta of 1937 was said to be good for 170kph (105mph), which might have been a tad optimistic. However, its bodywork was relatively lightweight at 126kg (340lb), which is what Touring's superleggera construction was all about. At any rate, the car - driven by Guidotti/Boratto - finished the gruelling 1000 mile event fourth overall.

While the racing version of the 6C 2300 B Mille Miglia Touring Berlinetta was an austerely finished competition car, the production model derived from it was given running-boards, wider doors and a higher roofline to accommodate rear seat passengers. It also had cutaway rear spats, semaphore arm indicators and plush seats.

You've seen those cooling slats around the radiator grille shield of a much more recent Alfa Romeo. This one had them first though, and it's the 6C 2300 B Mille Miglia four-seater Berlinetta housed at the Alfa Romeo museum at Arese.

Only 31 units of the long-wheelbase 8C 2900B were made, and ten of them were built by Touring of Milan with Berlinetta coachwork. It's demonstrably one of the finest cars ever to have been made, and this one is in the museum at Arese.

wheelbases, with uprated transmission and gearbox.

Between 1938 and 1939 198 cars in Corto form and 216 models in the Lungo version were made. There's a 6C 2300 B Corto Berlina finished in that delightful rich amaranto (Burgundy red) at the Alfa museum. The Corto Berlina looked very like the 6C 2300 Pescara model, but on the factory-bodied Lungo Berlina the spare wheels were housed prominently each side of the bonnet, cupped by the tapering ends of the front wings. The factory-bodied model was fitted with perforated steel disc wheels and chrome hubcaps, while Pinin Farina and Touring produced very much more elegant and forward-looking Berlinas on the

same platform. Both firms chose to hinge the doors on the central pillar, and Touring's car had the added sophistication of double hinges. Pinin Farina also produced a number of graceful cabriolets and coupés, while a Touring-built cabriolet was favoured by Benito Mussolini for his official parades. A limited run of 101 units of a 6C 2300 B Mille Miglia Berlinetta with streamlined bodywork was made in 1938, in celebration of Guidotti's fourth place in the 1937 Mille Miglia with a lightweight Touring-bodied 6C 2300 B. The 95bhp coupé was reckoned to be good for 170kph, though the vehicle offered to the public was rather different from the racer. The doors were widened for easier access and the

roofline raised at the back to provide headroom for two passengers in the rear seats, with a deeper rear three-quarter window to match. The spats over the rear wheels were missing, and metal wheel discs lent weight to a rather beefier-looking car. The silver example kept at Arese shows off details like the slats surrounding the radiator grille, which reappeared as a styling feature on the acclaimed 156 launched in 1997.

It's impossible to overlook the 8C 2900 B, even though it never existed in Berlina form, just Berlinetta and Spider. Derived from the victorious 1936 Mille Miglia cars run by the newly-formed Alfa Corse, the 8C 2900 B, these magnificent supercharged

straight-eight 2905cc machines developed 180bhp at 5200rpm, and were available in long or short chassis. The most memorable versions came from Touring of Milan, which made ten Berlinettas and at least one Spider. The Berlinetta's sloping rear end formed a bottom-hinged bootlid that opened up to reveal a pair of spare wheels. Two Lungo chassis are on show at Arese, both of which are quite well known; one is the aerodynamic 1938 Le Mans car. The 8C 2900 B has to be considered the company's flagship, and if Alfa buyers in general were not short of a bob or two, the owner of an 8C was particularly wealthy, since it cost twice the price of a 6C 2300 B Corto, which was 57,500 lira. Spider versions were made by Pinin Farina and his elder brother Giovanni Farina (the father of inaugural Formula 1 World Champion Nino, and patron of Stabilimenti Farina). There was also a very sleek factory-bodied Spider from 1941, nicknamed 'The Whale' at Portello, and possibly owned by King Michael of Romania, who also commissioned a much more formal Berlinetta version from Touring Superleggera.

The six-cylinder engine conceived by Jano back in 1925 metamorphosed from 1487cc through 1752cc up to 1917cc and then 2309cc, where it stayed from 1934 until 1939 when the bore was upped by 2mm to 72mm (stroke remained at 100mm) to give 2443cc. Logically, the new cars were identified as the 6C 2500 B. The factory Berlinas were powered by the revised engine, using a Weber 36 DCR carburettor. Externally they were very

This is the third series 6C 2500 Turismo which, on the face of it, looks pretty similar to the 6C 2300 B   Lungo with which it shared the 3250mm wheelbase. There were a few detail changes, such as the vents on the bonnet sides, bumper overriders and the Alfa badge on the base of the door.

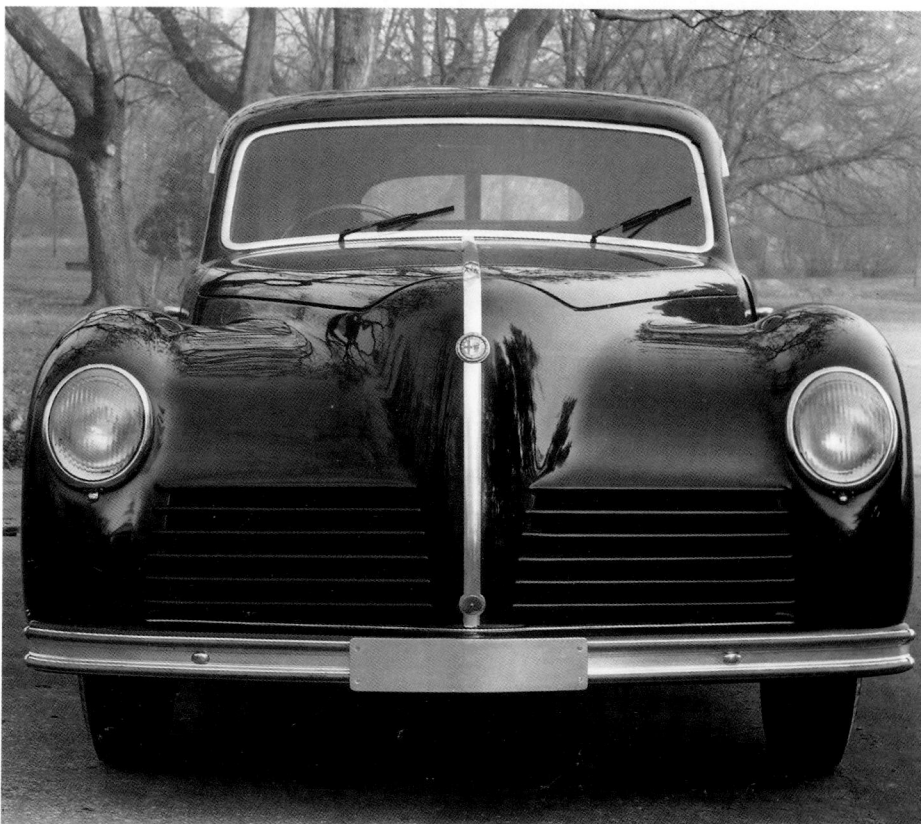

similar to the 6C 2300 B Turismo, Gran Turismo and Lungo models of two and three years earlier, apart from details like the front bumper, bonnet-side louvres and perforated pressed steel wheels. The five-seater Berlina was unveiled at the Berlin Motor Show in 1939, and later in the year the seven-seater model, with its spare wheels astride the bonnet, was introduced. Production continued into 1940

Whilst established Carrozzerie such as Touring and Castagna could be described as the conservative end of the spectrum, Pinin Farina was still producing rather more avant garde designs, such as this 6C 2500 Berlina of 1944. It has the two side grilles at the front, but lacks a central grille, the emphasis being placed on continuation of the bonnet centreline. The slim Alfa shield-shape grille that became one of the company's key postwar hallmarks was ushered in by Touring in 1941 on a handsome 8C 2900 B Berlinetta done for Prince Michael of Romania.

Unveiled at the Berlin Motor Show early in 1939, the 6C 2500 Turismo was marketed as a five-seater saloon. Apart from the larger bore size and detail changes - like the location of the wipers - it differed little from the 6C 2300 B. This example has steel disc wheels, fitted with the kind of ostentatious hubcaps normally reserved for contemporary spiders and Berlinettas. Note also the Portello coachworks badge at the base of the scuttle panel.

in spite of war breaking out in Europe, although, understandably, demand for cars of this type dwindled to just a handful by 1944.

The main alteration to the seven-seater 6C 2500 chassis was the switch from regular ladder-type cross members to an X-shaped diagonal cross brace. Apart from the Portello-bodied Berlinas, other examples came from Stabilimenti Farina and Pinin Farina, (one of which was constructed in 1943 for the German Field Marshall Von Kesselring), and Touring, whose Berlinetta was so large that it ought to qualify as a two-door Berlina. An interesting aside on this car is that it marked the first appearance of the narrow Alfa shield-shape radiator grille and paired side grilles, which would become the company's hallmark in the postwar era. Between 1939 and 1944, 282 five-seater Berlinas and 79 seven-seater models were made.

The competition derivative was the 6C 2500 Super Sport, bodied by Touring and assembled at Scuderia Ferrari at Modena, and typically identified by *Il Commendatóre* as the Tipo 256 - 2.5-litres, six-cylinders. Just seven chassis were verified as works-built Alfa

Romeo racing cars, and a further five units were built as racing cars in 1940, and fitted with Berlinetta coachwork by Touring. Another two roadsters were produced by Ferrari for the 1940 Mille Miglia under his Auto Avio Costruzione banner. The Super Sport chassis continued in production after the war, and I'd go so far as to say that these were about the most exciting looking racing cars ever to bear the Alfa badge.

By contrast the Alfa Romeo stable also included a number of commercial and utility vehicles. As well as light-duty trucks like the Autocarro Tipo 800 and heavy-duty haulage vehicles like the Tipo 85, from 1940 the company also made a sort of scout car called the *Coloniale* which was based on the 6C 2500 chassis. A cross between a Jeep and a regular Corto cabriolet, the rugged bodies were made by Castagna and kitted out with shovels, fuel cans and sundry other off-roading equipment. The Coloniale was intended for military use in the Italian colonies, and around 150 units were produced up to 1942.

In March 1939 the company's capital was abruptly increased to 240m from 30m lira, and its new name was

Società Anonima Alfa Romeo Milano Napoli. The Neapolitan reference concerned the aero engine plant at Pomigliano d'Arco, whose products powered, among others, the Savoia Marchetti SM75 record-breaking aircraft.

The 6C 2500 Sport Berlinetta, along with Spider and Cabriolet versions, was also shown at Berlin in 1939, and although 60 units were produced in 1940, only a few were made in the next few years. It was a key model, though, in that it transcended the war years and served to get production going again afterwards. Perhaps it should be regarded as surprising that Alfa managed to make any at all - like, two in 1945. But 58 units were made in 1946, and these were clad mostly with Pinin Farina and Touring Superleggera bodies, with Stabilimenti Farina, Boneschi and Ghia doing a few, and Castagna, Balbo, Graber and Monviso building one apiece up to 1950. The museum has a 1939 car on display, and it's largely in the headlight treatment that Touring's classic 6C 2500 Sport differs to the 8C 2900 B Berlinetta mentioned earlier as the lights are now almost completely incorporated into

This is a Touring-bodied 6C 2500 Sport Berlinetta, a crucial model because it transcended the Second World War. Launched in 1939 at Berlin, it surfaced again in 1946 and was made in relatively large numbers for a coachbuilt model - 1548 units, including Cabriolet and Spider versions - until 1950.

The 6C 2500 Sport Berlinetta by Touring was a natural successor to the 6C 2300 B Mille Miglia, although the later car had the rear-wheelarch spats seen on the competition Mille Miglia. Its beetle-back and streamlined tail were also very similar.

the rounded panels between wings and radiator grille. But, by 1945, cabriolets by Pinin Farina, Boneschi and Ghia all had the headlights positioned in the front of the wings in the most contemporary way.

Unfortunately, Italy was on the 'wrong' side in the Second World War, and, as the Allied thrust gained momentum, the Portello factory received direct hits in bombing raids in May 1943, and February, August and October 1944. The first raid destroyed two thirds of the plant; with the last raid, production came to a complete halt. As the war drew to a close, Wilfredo Ricart fled to sanctuary with Franco in Spain, where he began work on the Pegaso sports cars, whilst the paternalist Ing Gobbato was rightly cleared of complicity with the occupying Nazis, only to be gunned down outside the factory days later. The National Liberation Committee that exonerated Ing Gobbato now appointed Ing Pasquale Gallo as administrator, and, in July 1946, Dr Orazio Satta Puliga became head of Alfa's Design and R & D departments. The next phase of Alfa's history would see production exalted to a different level indeed.

Alfa Romeo Berlinas

3

REVIVAL OF FORTUNES

The Villa d'Este Superleggera Berlinetta won first prize at the Cernobbio Concours d'Elegance in September 1949.

After hostilities ceased, and thanks to assistance from the Marshall Aid programme, Alfa Romeo S.p.A. rose from the ashes of war, using components and materials salvaged from the wreckage. There was no assembly line as such, and the cars were simply built bit by bit and pushed from one station to another. Production started again very slowly in May 1945 with the 6C 2500 Sport Lungo, and the first postwar Berlina was a handsome round-roofed car designed by Castagna and built in-house by Alfa Romeo's re-emerging coachwork department. Apart from this car, and a big six-light Berlina by Boneschi that appeared in 1948, the majority of Alfas of this period were coupés or cabriolets.

Truck production began to get going again, too, and once again commercial vehicles and engines formed a significant part of the company's output. Models included the 4 x 2 A110 Autobus with concertina doors front and rear, the elegant 430 Autobus coach that featured a huge Alfa shield radiator grille as its front panel, and

there were even trolley-buses. The same 430 chassis served as a basis for light- to medium-duty trucks, panel vans and even horseboxes, whilst, at a totally domestic level, Portello also produced electric cookers. Hold the Rayburn, forget the Aga; the stove you wanted in your kitchen was an Alfa Romeo! They also made catering equipment, shutters, light industrial machine tools, aero engines (at Pomigliano d'Arco) and static diesel engines. Compare this utilitarian activity with the re-emergence of the fabulous Alfetta 158 racing cars, which began to assert their dormant prowess as early as 1946 in the hands of Varzi, Wimille and Trossi, and went on to carry Farina and Fangio to the World Championship titles in 1950 and 1951.

Column shift

One of the more remarkable 6C 2500 Sport coupés was the 1947 Victoria that came from Stabilimenti Farina, with a broad Americanesque chrome grille and a long fastback shape that almost prefigures the so-called Ferrari

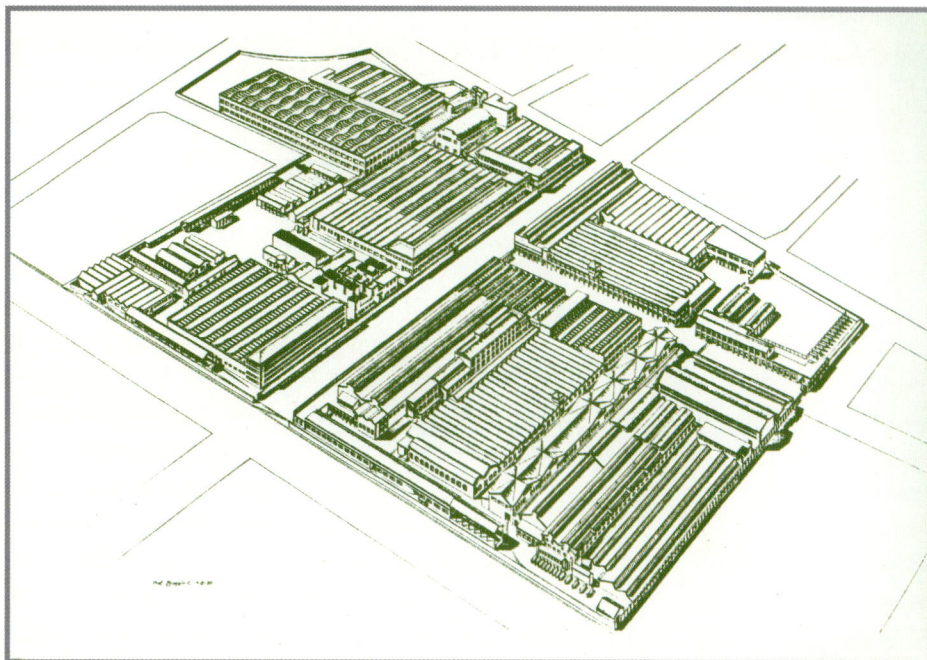

250 GT 'Breadvan' (by Neri and Bonachini) of 1962. Mechanically, postwar cars went over to column gearshift, in recognition of the prevailing fashion, particularly in the USA. Along with the column shift came revised dashboards that represented a major advance in civilising the driving environment, and ample front seats unimpeded by a floor-mounted gearlever could accommodate three people if necessary. The 6C 2500 B Sport of 1947 is best exemplified by Touring's Berlinetta models, featuring the tall, narrow Alfa grille and driving lights mounted between the headlights and grille. Wire-spoke or steel-disc wheels could be fitted, and by 1951 the body style included crease lines extending rearwards from front and back wheelarches, with single piece instead of split screen windscreen. Super Sport versions in Berlinetta form based on this chassis were made by Ghia, Stabilimenti Farina (drawn by Giovanni Michelotti), Pinin Farina and Touring, whose magnificent *Villa d'Este* coupé represented the zenith of big luxury grand touring Alfas. Styling now almost invariably featured integral wings, with very little definition between front and rear mudguards.

If the Villa d'Este was top postwar coupé, the *Freccia d'Oro* was top Berlina. Built on the 6C 2500 Sport rolling chassis with an interim length 3000mm wheelbase (the Villa d'Este measured 2700mm and the Sport Lungo Berlina wheelbase 3250mm), the Freccia d'Oro (Golden Arrow) was described by the company as a two-door Berlina rather than a Berlinetta, because of its ample proportions in the rear quarters. It featured a column gearchange, and ran on 17in rather than 18in diameter wire wheels. But, more significantly, the Freccia d'Oro broke new ground because its body was welded to the chassis, forming a single entity. This was the first time Alfa Romeo had done this with a production car, and it was an indication of the way things would soon be going. And despite its considerable mass, the Freccia d'Oro even competed with a measure of success in the Mille Miglia in 1948 (Fangio/Zanardi: third overall,) and 1949 (Rol: third overall,) and the 1949 Targa Florio, as well as that arduous Mexican road race, the *Carrera Panamericana* of 1950.

Company thinking had already turned towards a unit-construction monocoque body-chassis. In 1943,

Wilfredo Ricart had designed a five-seater Berlina prototype known as the *Gazzella*, powered by a six-cylinder, 2.0-litre engine with transaxle transmission and torsion bar suspension, and the project was revived in 1946. However, chief test driver Consalvo Sanesi declared it undriveable and, apparently on this advice alone, the project was shelved. This was a pity, because all the ingredients were right, especially the probable appearance of the curvaceous bodyshell. In the event, it was the Freccia d'Oro that was productionised, and 682 units were made up to 1951. From 1947 the majority were panelled by external coachbuilders, finished in a wide variety of colours and featuring full-length sunroofs, leather upholstery, carpeting and contemporary dashboard - still in righthand drive and with column shift.

An official function

For ministerial and ceremonial purposes, the 87bhp 6C 2500 Lungo chassis (3250mm) was adapted with the engine positioned on revised mounting points further forward in the frame, producing a much larger four-door passenger compartment. Of the 51 *Turismo con Motore Avanzo* chassis made, 32 were clad by Stabilimenti Farina between 1949 and 1951 and a further ten by Touring, which looked just like four-door versions of the Villa d'Este - essentially what they were. Now this was a really beautiful big saloon car. On the same chassis, Boneschi made a *Ministeriale* cabriolet in 1952, and Carrozzeria Viotti of Turin

The 6C 2500 Super Sport Villa d'Este Coupé (to give it its full name) was closely related in appearance to Touring's other creations based on the 6C 2500 chassis in Sport Berlinetta and Coupé forms. The postwar chassis was considerably strengthened and column shift allowed a third person to travel in the front.

The 6C 2500 Freccia d'Oro was described as a two-door Berlina, and was remarkable not only for its arresting presence, but also because it qualifies as the company's first attempt at productionising a monocoque. Rather than being bolted to the chassis, as was the norm, the body panels were welded to the chassis, providing a much tauter construction.

The Freccia d'Oro was always going to be an exclusive model, with just 682 units made between 1947 and 1952. The design evolved slightly: the slats on the side grilles became more prominent and the rubber inserts on the front bumpers visible here also expanded as overriders were fitted. Later cars also got driving lights in the bottom corners of the front wings.

There were no access problems with this luxury liner. It's a 6C 2500 Berlina quattroporte, created in 1947 by Pinin Farina on a Motore Avanzato chassis, and a harbinger of the rounded styling that characterised a whole generation of family saloons throughout the 1950s. Note the luxuriously appointed leather-clad interior and fabric covered roof.

Another variation on the 6C 2500 Turismo con Motore Avanzato is this Pinin Farina bodied model, fitted with a radio aerial. The frontal treatment is similar to that of Michelotti, who was employed at Stabilimenti Farina coachworks.

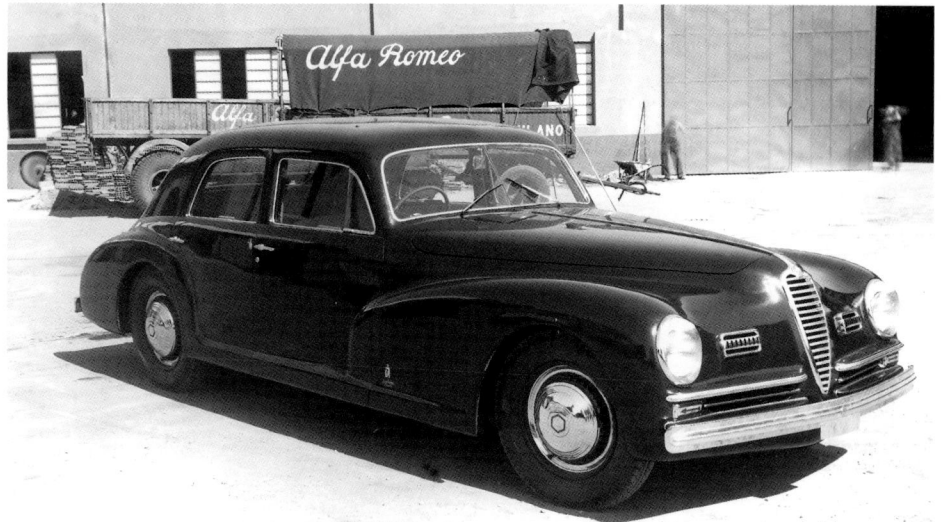

Centre left: Working to designs produced by the Portello coachworks drawing office, Castagna built this prototype for a four-door Berlina in 1947 on the 6C 2500 Sport Lungo chassis. Up to 1949, 110 chassis units were made.

Below left: This big Berlina is a 6C 2500 Turismo con Motore Avanzato - with the engine mounted further forward - one of 32 cars built by Stabilimenti Farina between 1949 and 1953. A total of 51 chassis were made and others bodied by Pinin Farina, Touring, Boneschi and Viotti, mainly intended for ceremonial and civic duties.

Right: This is a more conventional six-light Berlina built by Carrozzeria Boneschi in 1948 on a 6C 2500 Sport Lungo, the longer, wider chassis developed for these big saloons. The front windows have draught excluders, and the coachbuilder's badge is at the base of the front wing. Others to exploit this chassis were Pinin Farina, which made a two-door fastback Berlina for the King of Egypt, and Touring, which produced a 4/5-seater cabriolet.

In 1950, Alfa Romeo built its own Grand Touring body for the 6C 2500 Sport chassis, with a view to capturing the share of the Freccia d'Oro market that didn't care for the latter's round-back styling. But, despite installation of the 110bhp Super Sport motor, there weren't many takers, and only 145 units of the 6C 2500 Gran Turismo Berlina were built up to 1953.

made a small number of enchanting *Giardinetta* shooting brakes (Woodies) in 1950. The forward engine theme continued in 1950 using the 90bhp 6C 2500 Sport chassis (3000mm), and 151 units were produced with Pinin Farina-built four-door Berlina coachwork, the last one rolling out of the works in 1953. In 1950 one of these cars scooped the *Gran Premio d'Onore* at the Venice Lido concours, and, to an extent, the styling of this model heralds the advent of the 1900. It also bears some of the design hallmarks noticeable in other vehicles, such as the Bentley Mk VI, filtering down to the big Wolseley and Riley production saloons of the mid-'50s.

The final flowering of the 6C 2500 was a run of 150 units of the 3000mm wheelbase chassis, powered by the 110bhp Super Sport motor and built between 1950 and 1955. Like the Freccia d'Oro, the Berlina model was a two-door configuration, but instead of the huge arched back of the earlier model, the new Gran Turismo Berlina featured a distinctly coupé-like shape. By now the heavy old 6C 2500 was pretty much obsolete and was not a noted success in sales terms. The four-seater cabriolet version by Touring, which echoed most of the Villa d'Este's styling themes, proved more popular, even though far fewer were actually made. The concept of these superior cars built for an elitist clientele would endure in derivatives of the next generation of Alfas, the 1900 models.

It's a bit more stylish than a Series 1 Land-Rover, and that's because it's an Alfa Romeo. Just check the grille and there's that shield shape. The four-wheel drive Matta camionetta, officially known as the 1900 M (or AR 51/52), was in production from 1952 to 1957, with 2000 units going to the Italian military and a further 75 destined for civilian use. It was powered by a 65bhp version of the 1900 twin-cam, with a scavenge pump in the sump. A Matta Jeep took the military class honours in the 1952 Mille Miglia.

Monocoque construction

While the Alfetta 158s and 159s took the Grand Prix circuits of the world by storm - in the hands of Farina, Fangio and Fagioli - Alfa Romeo was busy making its Tipo 455 and Tipo 900 trucks in various formats - panel, dump and tanker bodies, as well as *Matta* Jeeps (any colour you like as long as it's green), fully-tracked bulldozers and snowploughs. But now mass-production had well and truly arrived; in the Portello car division great metal presses turned out the panels for the new 1900. Although the new all-in-one body-chassis construction broke with the traditional separate chassis and bodywork fabrication, fundamental links with the old powertrain were carried over into the new 1884cc, 80bhp, four-cylinder twin-cam engine. In commercial terms, the new build

As if the curvaceous lines of the 1900 Berlina weren't attractive enough, the bonnet also featured this elaborate mascot.

methods meant more economical production, hence lower prices, so more people could afford the cars. The assembly line was as yet fairly basic and labour intensive, with bodyshells moving along tracks aboard simple trolleys. Preparation for painting was carried out by hand, with any necessary touching-up done with a spray gun on the completed cars. The drivetrain was fitted by three operators using an overhead hoist, with one man on each side and one below, with the fully trimmed cars aloft on an elevated track - hence the likelihood of the car sustaining paint damage during this process.

Development work had been long and painstaking, carried out under the direction of Dr Satta Puliga and mechanical engineer Giuseppe Busso, and testers

As the company went over to mass-production processes in 1950, we see the new 1900 model being given a final rub-down before entering the spray booth in the Portello paint shop. Painted shells emerged on the left.

Painted 1900 shells turn a corner on the assembly line. Each one is mounted on a wheeled frame and guided along a rudimentary track as trim and mechanicals are fitted. The atmosphere in the light and spacious assembly area appears calm and unhurried.

Giambattista Guidotti and Sanesi. The elegant curves of the 1900's monocoque bodywork with its 2630mm wheelbase were drawn up under the supervision of bodyshop head, Ivo Colucci, and metamorphosed through several evolutions before the definitive shape was arrived at. The frontal treatment of the first prototypes was very similar to that of the Freccia d'Oro, although the rest of the car was all new, with slab sides and rounded roof, wings and bootlid. In prototype form, the rear quarters fell away quite steeply until fairly late in the day, and the model was officially launched in 1950 at the Grand Palais that staged the Paris Show. For the most part, 1900s were painted in fairly sombre colours - five varying shades of light and dark grey, plus deep cobalt blue and a lighter Capri blue.

The 1900's engine consisted of an aluminium head, rocker covers,

The regular 1900 Berlina Normale was in production between 1950 and 1954, during which time 7407 units were made. Almost predictably, it spawned a host of 2+2 coupés and cabriolets of varying degrees of attractiveness, from Castagna, Touring, Pinin Farina, Colli, Ghia and Boneschi, based on the 1900 L chassis floorpan.

bellhousing and rear section of the gearbox casing, with an iron block, crankcase and main gearbox casing. The rocker covers and head were painted crackle finish matt black. Instead of the regular Weber 40 DCF5 and 40 DCL5 or Solex 33 PBIC and 40 PAI carburettors, the prototypes used SUs, which were easier to adjust. Up-

holstery in the production cars was less opulent than before, with cloth upholstery and rubber matting being the order of the day. Like most Berlinas, the 1900 was endowed with a spacious boot, and the spare wheel was located vertically inside the right-hand inner rear wing. The advantage of this was that it could be removed fairly easily in

The 1900 was powered by a new 1884cc four-cylinder twin-cam engine, which, when fitted with the Weber 40 DCF 5 or Solex 40 PAI carbs, produced 85bhp at 5200rpm. The cylindrical air filter, absent here, was angled towards the rear of the engine. The cylinder head and rocker covers were cast aluminium, as were the bellhousing and rear section of the gearbox casing, while the block was cast iron.

1900 T.I. also had fatter stub axles, larger drum brakes at the front (which occupied almost the entire wheel casting), and bigger brake cylinders. A special air intake was fitted for cooling each front brake. Wheel sizes broadened from 16in to 16.5in diameter to accommodate the new 165 x 400 radial tyres. The justification for this higher-specification model was to field a competitor to Lancia's *Aurelia*, and a contender for the Turismo Internazionale

the event of a puncture without having to unload all the luggage. Another significant innovation ushered in with the 1900 was the switch to left-hand drive, the implication of which was that, in future, right-hand drive vehicles would be made only for export to appropriate markets.

Front suspension was independent by upper and lower wishbones, telescopic dampers and coil springs. The 1900's rear suspension was more basic than its predecessors', with a live rear axle located by trailing arms, hydraulic dampers and coil springs and, on the first 1400 units, a transverse tubular arm, which was superseded by a triangular radius arm in 1952 in a move to counteract undesirable oscillations in the rear suspension. Twin-shoe finned-aluminium drum brakes were fitted front and rear, and the 4in-wide wheels were pierced with oval holes to assist brake cooling. The 1900 *Berlina Normale* was in production from 1950 to 1954, with 7407 units built.

The T.I. variant (which stood for *Turismo Internazionale*) introduced in 1953 was equipped with a pair of twin-choke Solex 40PII carbs, or Weber 40 DCO3s if required, breathing through an air intake unit designed to ram air into the carbs at high speed and pro-vide a sort of supercharging effect. An efficient four-branch exhaust manifold was standard issue, merging into two exhaust pipes, and the 100bhp

By 1953, Alfa Romeo's 1900 had come under threat in competition circles from the Lancia Aurelia, and the company responded by bringing out the 100bhp Turismo Internazionale version. It was the first production car to be fitted with two twin-choke carbs - Solex 40 PII or Weber DCO3 - and the air intake was stretched to the front of the car for ram-air effect. The 1900 T.I. was only around for a year, so 598 units isn't really such a small tally.

During its final months of production, the 1900 Super was finished in the popular two-tone colour scheme. This was probably inspired by the short run of two-door 1900 Berlinas created by Boano between 1955 and 1957. On the 1900 Super, the effect of the two-tone scheme was exaggerated by chrome strips that even went so far as to suggest that the front wing moulding extended half-way along the front door.

racing category. Just 598 T.I.s were made between 1953 and 1954.

An oddball variant was the 1900 *Berlina Pantera*, finished for the Italian police with uprated suspension and bullet-proof two-panel windscreen, as well as bullet-proof mesh guards ahead of the front wheels for protection in high-speed car chases. The 1900 T.I. and T.I. Super was used as a competition car, initially in a *Raid* through the Sahara desert to Somalia, going on to feature in events like the 1952 Alpine Rally, the 1953 *Tour de France* and Tour of Sicily, and the 1954 *Giro d'Italia*. A T.I. Super won the Turismo category in the 1954 Tour de France as well as the Mille Miglia, while a T.I. in the hands of Corini/ Artesoni won the Tour of Sicily outright that year. And a six-car team led by Consalvo Sanesi contested the 1954 Carrera Panamericana.

The 1900 Super was introduced in 1954, with the engine now bored out to 1975cc, a new crankcase and fed by a single Solex 40 PAI carburettor. The following year duplex silent chains were adopted rather than the gear and chain system used previously, and the camshafts now rotated in the opposite direction, implying a new firing order - 1-2-4-3 instead of 1-3-4-2. In accord with the times, the Super was given more chrome trim, including bumper overriders and a line that ran from sidelight to sidelight across the side-grilles. New rear-light clusters were fitted, which included a brake light and indicator. In the cabin, there was a front bench seat with central armrest with two-tone insets, and the dashboard was smartened up with three circular dials, with rev-counter to the right and speedo to the left. The steering wheel was also given the two-tone treatment with cream bakelite at the hands-on twenty-to-four position. This popular mid-'50s finish was applied to the second series 1900 Berlinas, along

with yet more chrome trim lines, and they were known as the *Primavera* (Spring) model. Two-door versions were produced by Carrozzeria Ghia and by Carrozzeria Boano - and Boano made 281 units of the two-door Primavera between 1955 and 1957. To complicate matters a little, the 1900 Super could also be fitted with T.I. specification ancillaries to become the 115bhp T.I. Super, making it one of the fastest production saloons in the world. Some 485 T.I. Supers were made between 1954 and 1957; by 1959, when the last 1900 left the factory, 8512 units of the 1900 Super had been built.

Coachbuilt coupés

The factory-built Berlina was accompanied by a host of fairly similar coupés made by the leading coachbuilders, including Touring, Castagna, Pinin Farina, Vignale, Ghia, Zagato and the Vercelli-based Carrozzeria Francis Lombardi. The 1900 C Sprint and hotter

Portello's model line-up in 1954 consisted of the ubiquitous Romeo van, the 1900 C Super Sprint by Touring, the 1900 Super and the Giulietta Sprint.

Super Sprint drawn by Aquilino Gilardi for Touring Superleggera was undoubtedly the classic 1900 coupé, and was the one endorsed by the factory, while Michelotti's design for Vignale was perhaps the most futuristic. Castagna's were all aluminium, while Zagato's version was raced extensively by Sanesi. But quite the most outrageous were the three Berlina Aerodinamica Tecnica (B.A.T.) cars done by Franco Scaglione for Bertone. It's worth noting that the second series 1900 C coupé, which Alfa Romeo commissioned from Touring in 1956, was a much leaner car that, in frontal aspect, bore a strong family resemblance to Bertone's Giulietta Sprint.

Two-door 1900 Berlina Normales were produced by Bertone and Boneschi, and cabriolet versions were made by Stabilimenti Farina and Pinin Farina, whilst Boneschi's finned Astral cabriolet went off at a seriously space-age tangent. A lengthened rolling chassis designated the 1900 L was available to specialist coachbuilders, with the same dimensions and wheelbase as the standard car. An example of this was built up as a three-light saloon in 1952 by Carozzeria Colli, which was possibly better known for contemporary work on the bodyshells of the dramatic 3000 CM sports racing coupés - as driven by Fangio in the 1953 Mille Miglia and in Spider form at the Supercortemaggiore Grand Prix at

Merano. Colli also made a chunky plus-two 1900 coupé that could just as easily qualify as a Berlinetta. The AR 51 and AR 52 four-wheel drive 1900 Matta jeep was something else again, having been commissioned by the Ministry of Defence as an all-terrain, general purpose vehicle, and 2500 units were made between 1952 and 1957. It was used not only by the Italian military but also the fire service and for snow clearance.

Worthy of a mention here is a prototype AR 52 *Matta Giardinetta*, which looked a bit like an early Toyota Land-Cruiser. If productionised it would have pre-dated the burgeoning off-roader market of the 1990s by some four decades ...

4
PRETTY
WOMAN

One of the prettiest cars ever made: the Giulietta Sprint, introduced in 1954, which helped whet the appetite of motoring enthusiasts for the forthcoming Berlina, whose launch it pre-dated by a full year. The 750-series Giulietta Sprint was styled by Franco Scaglione and Nuccio Bertone, and, with certain stylistic and mechanical updates, it was built at Bertone's small factory until 1963. In the final year of production it was known as the Giulia Sprint, and fitted with the 1570cc engine from the incoming 105-series cars.

It's passed into legend that the Giulietta name originated at a high-ranking dinner party during the launch of the 1900, when a Russian prince remarked that there were plenty of Romeos at the Alfa table but not a single Giulietta. True or not, the Sprint model that introduced the range was unquestionably one of the prettiest cars ever made - and that's still true today.

The Sprint was introduced at the Turin Show in 1954 in the wake of a well-documented national lottery scam, and was powered by a new, all-aluminium 1290cc twin-cam engine, which was unheard of in a production car at this level. The prototype even had a rear hatch incorporating the back window, a feature that didn't extend to the production model. At this point in time, the Berlina was still at the prototype stage. Some original renderings for the Sprint were done by

Distinctive frontal treatment of the Giulietta Berlina Normale. The car's specification was unprecedented in that it made available to the general public a level of performance and handling normally reserved for competition cars due to the light-alloy high-revving engine, wishbone front suspension and synchromesh gearbox, along with an efficient drum brake system.

Mario Boano when he worked for Ghia up to 1952. Ghia hadn't the capacity to take on the Sprint project so it went to Nuccio Bertone, whose own ideas were brushed up by Franco Scaglione to produce the final design. The Giulietta Sprint went into production in Bertone's cramped workshops - subsequently enlarged - and metamorphosed through 750-series model to 101-series model in 1959, with certain cosmetic changes: when the 1600cc twin-cam replaced the 1300cc engine, the Sprint and its Pininfarina-built Spider sibling became known as Giulias. The Sprint coupé was always largely hand-built, whilst the Portello-made Giulietta Berlina was created along a recognisably more modern production line. If you went to Longbridge

Side-rails supporting the Giulietta floorpan are welded at an early stage in the Berlina build at Portello. Other components are stacked up to the rear, awaiting fitment.

By 1955 the Giulietta production process was more assured than it was with the 1900, and the monocoque bodyshell was built up with panels clamped prior to welding together in assembly jigs. As you can see, they were mounted on a platform that moved along a track, which is itself raised up on a wooden dais.

49

in the 1990s you'd find the Mini was produced in much the same way.

Constructed entirely at Portello, metal presses fabricated the panels that were welded together to form the Berlina's monocoque body-chassis units. The panels that combined to form the body-chassis unit were welded together in a series of jigs mounted on wheeled benches which travelled along a track elevated on a raised wooden platform. The completed units went down the line mounted on heavy-duty trolleys, as operators ground off excess metal from the roof seams. Any imperfections in the bodyshells were attended to at this point and filled, and with bodyshells suspended from an aerial gantry the cars were hosed down prior to being painted. After passing through the drying tunnel, the painted shells progressed via overhead hoists to mechanical mounting stations where all trim was fitted. It was only then, with front wings and panels protected by sheets, that the engines and ancillaries were installed. Up on overhead piers, the Berlinas were finished off in parallel with Sprints and Spiders, although they outnumbered them considerably.

So, the four-door Berlina had been in the offing for at least three years, judging by factory coachwork assembly drawings from 1952 which demonstrate that the Berlina in its final incarnation was hardly changed from the original concept - a scaled down 1900, in effect. The project was driven by Dr Orazio Satta and Ing Giuseppe Busso, and the design given its final tweaks by Giuseppe Scarnati in 1955. The car was launched later that year at

A member of the Portello workforce installs the window winder mechanism in a Giulietta Berlina as the painted cars go through the glazing stage. The fresh paintwork is protected by shrouds.

Broadly, the Giulietta Berlina had the same rounded lines as the larger 1900 model, but in more compressed form. This car is finished in the distinctive coral red colour.

the important Turin Show.

Despite innocuous rotund styling, it was an amazing product to spring on the family car market in that its mechanical specification was directly based on far more exalted competition machinery. At its heart beat an all-aluminium, 1290cc, twin-cam engine with hemispherical combustion chambers, domed pistons and centrally located spark plugs. This was allied to a synchromesh gearbox, with an aluminium differential casing, and the monocoque chassis rode on twin wishbones, coil springs and dampers up front, with two radius arms and a triangulating link that located the solid rear axle, further supported by coil springs and dampers. Carburation was by Solex C32 BIC, later supplanted by the Solex C32 PBIC. Inside, the Giulietta Berlina was rather austere, with ribbed hound's tooth check upholstery and wool and rubber floormats.

The following year power output was lifted from 50bhp to 53bhp, and the gearbox went over to Porsche-type synchromesh, followed by a new worm and roller steering box. Unlike its Sprint and Spider siblings, which received the *Veloce* treatment in 1956, the Berlina was ignored in this respect. Veloce modifications were instigated in order to defeat the opposition - typically the Porsche 356 (and subsequently the Lotus Elite and Simca-Abarth), and included a higher compression ratio, higher lift cams, twin Weber carburettors instead of the sin-

The 750-series Giulietta range was powered by the 1290cc aluminium twin-cam engine, producing 50bhp initially, and in this case retro-fitted with a pair of open-trumpeted Weber twin choke carburettors.

The Berlina T.I.s may not have had such an illustrious competition career as their Sprint-based siblings like this Zagato-bodied Giulietta SVZ of Bart Baljon, but it's interesting to reflect that they were produced against a background of widespread success in European GT racing in the late '50s and early-60s in events such as the Mille Miglia and Tour de Corse, which all helped increase the marque's esteem.

When the 101-series Giulietta range was introduced at Frankfurt in 1959, modifications to the engine included relocation of the fuel pump to the crankcase, and a larger cylinder head incorporating larger diameter valve stems and bigger diameter main bearings. By 1962 a six-blade nylon cooling fan was used, and this car has a later 105-series air filter system fitted.

gle Solex, tubular exhaust manifolds and a finned cast-aluminium sump that contained a built-in oil cooler and surge baffles to aid temperature control.

Instead of introducing a Veloce version of the Berlina Normale, Alfa

Romeo followed the route taken with the 1900 and brought out the Turismo Internazionale in 1957. Despite retaining only a single twin-choke Solex carburettor, the 65bhp Giulietta Berlina T.I. proved almost as quick as the Sprint during tests at Monza; again, this was something of a revelation at the time for a 1300cc saloon car. Apart from having a rev counter, the T.I.'s specification and state of trim was otherwise no different to the regular Normale.

Until 1959, the only alteration to the Berlina's exterior was the fitting of

The Giulietta SZ was built by Zagato on the Sprint Speciale floorpan, essentially for competition purposes, and 169 units were made between 1959 and 1963. The model's potential was demonstrated by its 10th place overall in the 1962 Le Mans 24-Hours, and 7th overall in the Targa Florio the same year.

In its final evolution, introduced at Frankfurt in 1961, the Giulietta T.I.'s side grilles were embellished with much chrome, which tended to dominate the central Alfa shield. This car belongs to John Brittan.

From 1961, the Berlina had floor shift - optional in the Normale and standard in the T.I. model - along with reclining front seats and a linear speedometer flanked by rev-counter to the left and water and oil temperature gauges to the right. Brake and clutch pedals were floor-hinged, and accelerator pedal top-hinged.

larger asymmetrical headlights, but at the Frankfurt Show that year the facelifted model was unveiled with revised side-grille treatment (like the Sprint), rear wings that had almost grown fins, headlights with prominent bezels, and rubber-faced overriders on the bumpers. The T.I. gained side indicators in the front wings, while the bonnet emblem (or handle) that graced the earlier cars disappeared. Inside, the dashboard was uprated and interior trim mildly reworked. Engine internals were beefed-up slightly with enlarged valve stems and crankshaft main bearings.

Along with the Sprint coupé and Spider, chassis numbering began with the digits 101- at this point, and the Berlina's designation switched from 750-series to 101-series. A six blade plastic fan was fitted in 1961 as the Normale's power rose to 62bhp, and at the Frankfurt Show that year the final styling alterations extended to adjustments to the bonnet and boot openings. The T.I. was fitted with a Solex C32 PAIA-5 carb, and its power output rose to 74bhp, so that it was capable of a 155kph maximum speed.

While the Giulietta Sprint and its SVZ and SZ derivatives (built by Zagato on Sprint Speciale floorpans) were Alfa's principal competition cars during the late '50s and early '60s, the T.I. also featured in domestic events like the Giro d'Italia in the Turismo class, and there were numerous successes in private hands. Top French rallyists Consten/Hérbert took a class win in the 1958 Tour de France, followed by Masoero/Oreiller, while Masoero made fastest time of day at the Mont Ventoux Hillclimb ahead of Baillie's Jaguar. Only 25 cars finished the 1958 Alpine Rally, and Giulietta T.I.s were the first three cars home, while another took the ladies' prize in the Tour of Corsica that year (Aumas/Wagner). The 1960 Geneva Rally was won by Oreiller/De Lageneste in a T.I., endorsing the model's versatility and keeping the Berlina flag flying while the contemporary coupé versions dominated their own classes. The most significant Giulietta T.I.-based partnership in European rallying in the early 1960s was probably that of Masoero/Oreiller, with Masoero/Maurin in Giulia T.I. Super from 1963, when they were fourth overall in the Tour de France and fifth in class the following year. This pairing then went over to running GTA and GTZ.

I recommend the excellent book *Veloce - The Racing Giuliettas* by Donald Hughes and Vito Witting da Prato for the detailed story of the Sprint Speciale, SVZ and SZ competition history. We're jumping ahead a bit now, and it's a reflection of my age that the first time I saw a Giulietta Berlina driven in earnest was at the Snetterton 500kms on Good Friday 1966. This was the event that posed the titanic battle between the Autodelta GTAs of de Adamich, Rindt *et al* with the Lotus Cortinas of Whitmore and Stewart amid frequent torrential downpours, while hoards of tiny Fiat Abarths scampered

Yes, believe it or not, this car bears an Alfa Romeo badge, albeit in tiny script on the front wing. Between 1959 and 1964, Alfa made more than 70,000 Dauphines and Ondines at Portello under licence from Renault.

at the tail, snapping at the Mini-Coopers. Also towards the back of the field circulated a lone white Giulietta Berlina, taking the long Coram right-hander more or less on its door handles. Those were the days, eh!

Final evolution

In 1961, the 100,000th Giulietta rolled off the Portello assembly line, and the final evolution of the Berlina took place in 1962. Giulietta Berlina production finally amounted to a total of 128,822 vehicles. The most prominent external change to the end-of-line model involved the much larger chrome grilles applied to the front of the car, which totally lacked the delicacy of the original yet were in keeping with the times. The Sprint had received the same treatment in 1959, but volumes were smaller and

Only produced as a prototype, the Tipo 103 was built in 1960 to assess the possibilities of a front-wheel drive layout. Its transverse-mounted 896cc alloy twin-cam - the smallest configuration of this unit - coincided with the Mini, although the Alfa was a four-door model and the proportions were quite different.

changes easier to institute. The Berlina now got separate reclining front seats, a linear speedo, and more importantly, a floor shift, which was optional in the Normale and standard in the T.I. For some reason UK-spec cars were accorded the floor shift a year earlier.

Variations on the Giulietta Berlina theme included a stretch limo by Carrozzeria Colli, with reinforced floorpan and specially strengthened wheels, which was produced for ceremonial duties. A more practical car was the Giulietta Promiscua estate, in which the station wagon back end was integrated very nicely with the Berlina body. Carrozzeria Colli built 72 of these as 750-series and 19 as 101-series cars. Perhaps surprisingly, there was quite a long waiting list for delivery of a Berlina and, to avoid this, wealthier clients went to the specialist Carrozzeria, which made what now look like kitched-up variations on the factory Berlina theme - lots of chrome trim and duo-tone paint schemes. Among these were Zagato, Vignale and Moretti of Turin.

Alfa's commercial vehicle production in the mid-'50s now ranged from the ubiquitous Romeo van to the Mille medium truck and the Sicar and Turbocar coaches, with engines for heavy equipment, including excavators, road-rollers and naval harbour launches, also prominent. The workforce totalled 10,000 in 1958, and the company re-acquired land at Pomigliano d'Arco near Naples, as well as the Livorno-based Spica fuel injection concern. By 1960, work had begun on a new factory on the new greenfield site at Arese, and Ing Pasquale Gallo retired, to be replaced as Chairman by motor racing enthusiast Dot. Giuseppe Luraghi.

Between 1959 and 1964, Alfa Romeo also assembled Renault Dauphines under licence, and, somewhat bizarrely, the little 845cc rear-engined French *bolide* sported Alfa badges. It has an Alfa type number - tipo AR 1090 - so qualifies as an Alfa saloon in the same way as does the Nissan-based Arna of the 1980s. Having the engine hung out behind the rear wheels was not a problem for Italians reared on baby Fiats with the same configuration, and the Dauphine's front suspension at any rate was state of the art, consisting of double wishbones, coil springs and dampers allied to disc brakes. For about a year between 1961 and 1962, the upmarket Dauphine model known as the Ondine was produced at Portello; from 1962 to 1964 at Pomigliano d'Arco was also made - wait for it - the French farmer's delight, the Renault 4 utility car, but this model didn't carry any Alfa badges. They were made in serious volumes too, for the period. Production of Dauphines and Ondines totalled 70,502 units, and 41,809 Renault 4s were made in two years. Also from 1958, Alfa Romeos were assembled from CKD (completely knocked down) kits imported to South Africa, by Titan Industrial Corporation in Cape Town. A legacy of this operation (or rather the climate there) is that, even today, it's still just about possible to find rust-free classic cars built there.

Mini rival

It's worth including a prototype here known as the Tipo 103. This was a conceptually dramatic front-wheel drive saloon, built in 1960 and powered by the smallest configuration - at just 896cc - of the all-alloy twin-cam engine. It was therefore similar in engine size to the 848cc Mini, but nearly two feet longer, some six inches narrower and two inches lower than that British paragon of compact design. So it was quite different proportionally and with four-doors and a boot, whose lid contours prefigured that of the Giulia T.I. As a space saving measure, the spare wheel was housed under the rear seat, which, in the event, could have posed some practical difficulties. Suspension was by wishbones, coil springs and trailing arms - like the Giulietta. More radically, the Tipo 103's all-synchromesh, four-speed gearbox was constructed as part of the engine and, like the Mini, the whole assembly was transverse-mounted. It would be another 29 years before the twin-cam was orientated on the east-west axis, with the 164 TwinSpark model of 1989. Unusually for 1960, though, the camshafts of the Tipo 103 prototype were driven by synthetic rubber belts, while the generator used a belt and intermediate pulleys. The crossflow engine developed 52bhp at 5500rpm, which was better than all the Minis except for the later Cooper S models. The downside was that the Tipo 103 weighed in at 1600lbs, so top speed was a disappointing 80mph.

Although there were some stylistic similarities with the much bigger 102-series 2000 saloon and the forthcoming Giulia T.I., the project never got off the ground due to the cost of erecting another production line at the new Arese plant. There was also Alfa's lucrative contract with Renault to bear in mind, as the Tipo 103 would have been direct competition for cars like the R8. So, just a single Tipo 103 was built, along with three engines, and it can be seen in the museum at Arese. The project was re-examined when the Alfasud was under consideration, but by the late 1960s much of its design was obsolete.

5

BIG SISTERS

Viewed head-on, the big 102-series 2000 saloon was nothing if not stylish, and was characterised by a raised bonnet moulding that incorporated the Alfa shield, complemented by lavish quantities of brightwork. This is a Sao Paulo-built car, made in 1961, which belongs to Michael Swoboda and his son Leandro. A true family Alfa.

Just as the 1900 was an appreciably bigger car than the Giulietta, so too was the 2000 that appeared in 1958. This model has to be seen as a reworking of the 1900 theme; a reasonably large car in what we would now call the executive class, the province in recent years of the 164 and 166. The concept was the same - a three-box, four-door saloon in typical '50s fashion with bags of chrome trim along the sides and around the window frames, and a grille layout that was, to an extent, taken up in the facelifted Giulietta Berlina of 1961.

The 102-series 2000 was powered

Some of the ancillaries on this four-cylinder 2000 engine have seen better days. Notice the cylindrical air filter, angled rocker covers and enormous brake servo.

57

Compared with the 2000 model, the 2600 Berlina was relatively restrained, especially in its later form, with the chrome strips that had previously adorned its flanks now absent.

by the 1975cc engine used by the old 1900 Super, allied to a five-speed gearbox with Porsche-type synchromesh, but equipped with a fresh cylinder head and Giulietta-derived valve gear with bucket tappets. It was fed by a Solex C35 APAI-G carburettor, as used by the Giulietta T.I., albeit with different settings. The model was in production until 1962, during which time 2893 units were made.

Alongside the 2000 Berlina was the Touring-built 2000 Spider, made until 1961 with 3459 units built, 40 of which were by Vignale. The Spider was constructed by Touring as a two-seater and latterly as a two-plus-two. A 2000 coupé with twin headlights was made by Bertone between 1961 and 1962 (708 units) on a slightly longer wheelbase, featuring plush leather upholstery and controls taken from the Giulietta SS and SZ. The design, attributed to the young Giorgetto

One variation on the 2000 Sprint theme was this concours-winning Coupé, bodied in 1960 by Carrozzeria Alfredo Vignale and belonging to Delmas Greene.

Giugiaro, anticipated the much more widely seen Giulia Sprint GT introduced in 1963.

Straight Six

If we seem to have passed over the 2000 Berlina rather swiftly, that's because it wasn't so different in specification to the 1900. But the 2600

Berlina - designated 106-series and launched at Geneva in 1962 - was a rather different proposition. Although the design of the body was largely unchanged apart from the second pair of headlights mounted on the grille, it was powered by a new all-aluminium straight six with seven main bearings, removable cylinder liners and Giulietta-

When it came out in 1962, the 2600 Berlina featured as much chrome trim as its predecessor, but lacked the entertaining bonnet detail. This car, belonging to Lorenzo Quevedo Negrete, is also fitted with a pair of spotlights.

Below right: Big car, this. Viewed from the side, the 2600 Berlina has the proportions of saloons such as the Pininfarina-styled Austin Westminster. Yet it also carries something of the raked windscreen and wraparound rear window of the contemporary Giulia T.I.

like hemispherical combustion chambers. It developed 130bhp and was fed by twin Solex 32 PAIA-4 carbs, and from 1964 when the design was mildly facelifted, a Bosch alternator replaced the old dynamo electrics. This uprating included installation of bucket seats in the front and an electric clock, and elimination of some of the chrome trim of the original '50s design served to update the 2600, although it subsequently lacked the idiosyncratic charm of the old 2000. As befitted its intended role, the cabin of the 2600 Berlina was relatively opulent and was, of course, far more spacious than either the Bertone Coupé or Touring Spider. Unlike the Giulietta Berlina, there was ample accommodation in the rear as well. Boot space was equally generous. It was normal for left-hand drive cars to have a column gear shift, and those with right-hand drive to have a floor shift, with the brake and clutch pedal floor-hinged and the organ pedal accelerator top-hinged.

Shared powertrain

As with the 102-series 2000 models, the 2600 Berlina had similar Bertone Coupé and Touring Spider siblings. All three models shared the same powertrain and mechanicals although the sports models were fitted with three twin-choke Solex carbs rather than two, and thus with appropriate

Dashboard of a right-hand drive 2600 Berlina, in which everything has its place and yet the overall impression is somewhat discordant; perhaps because there are so many different shapes. The instrument panel is dominated by a linear speedo, while the steering wheel has a chrome horn ring typical of the period.

An early Giorgetto Giugiaro design: the handsome 2600 Sprint was built by Bertone on an updated 1900 floorpan, and was produced between 1962 and 1966.

Compared with the Bertone-designed Sprint model, far fewer 2600 Berlinas were made, despite a lengthy production period. Even the Touring-bodied Spider version managed a higher build quantity. This car belongs to Mike Abthorpe.

manifolding each cylinder was individually fuelled. The 2600 Bertone Coupé was made from 1962 to 1966, and differed from the 2000 version only in having a bonnet air intake, although interior trim was altered slightly in 1965. The 2600 Spider featured the same changes to its façade as the Sprint, with an extra set of headlights and a bonnet air intake.

Pininfarina also came out with prototypes based on the Coupé and Spider, while Zagato's steel-bodied 2600 SZ made from 1965 to 1968 is a rare example of how a specialist coachbuilder can successfully reinterpret a recognised design.

An exclusive 106-series design that was both light and graceful - and which didn't have the baggage of the '50s 2000 precedent to encumber it - was a Berlina by Carrozzeria O.S.I., known as the 2600 De Luxe. Turin-based O.S.I. (*Officina Stampaggi Industriali*) made the 2600 De Luxe in only very small numbers - 51 units in total - but the specification was quite exclusive and ahead of its time for 1965, including air conditioning, leather upholstery and the three-carburettor engine, which endowed it with 145bhp. However, in the grand scheme of things, the 2600 was not a large volume model. Just 2051 factory Berlinas were made - less, you'll note, than the 2000 version, but a much greater number of 6998 Bertone Sprints was produced, plus the 103 Zagato SZs. By comparison, 2257 Touring Spiders were built, and this short, but by no means insignificant, episode in the company history drew to a close in 1968.

Also powered by the 2584cc straight six twin-cam, a small number of the 2600 De Luxe were made by Carrozzeria O.S.I. in 1965. This example is missing its Alfa grille shield and front bumper, unfortunately.

The coachbuilder Zagato was more commonly known for his stripped-down racing versions of Alfa Romeo sports models, but his rendering of the 2600 Sprint contained a surprisingly luxurious interior. The steel-bodied 2600 SZ also featured an exaggerated Alfa shield and ovoid headlights behind plexiglass fairings, plus a typical coda tronca Kamm-tail at the rear.

6

BELLA GIULIA

The Giulia saloon range that was around for much of the 1960s was introduced by the Giulia T.I. Unlike the Giulietta Berlina that preceded it, there wasn't a Normale *version; Alfa went straight ahead with the Turismo Internazionale with its sporting connotations.*

The long-running 105-series Alfa range was ushered in by the Giulia T.I. Berlina, launched at Monza in June 1962. The essential elements of the Giulietta powertrain were retained, but conceptually, the Giulia represented a major move towards rationalised and (relatively) low-maintenance engineering. As classic Alfa fans know only too well, these cars function properly only with rigorous servicing, but the Giulia was a step in the right direction as at least the suspension and steering systems no longer required lubrication. In this sense, some of the Giulietta's thoroughbred pedigree was lost, but the longevity of the Giulia saloon's production run - from 1962 to 1974 (the mechanicals endured to 1993 in the 105-series Spider) - is testament to the correctness of the formula.

The new model was designed by a team known by the tortured acronym DIPRECAR (Direction of Experimental Bodywork Project), with Satta at the helm and prime mover Ivo Colucci assisted by Antonio Moneta, Giorgio Pavesi and Renzo Sfondrini. Model makers Giuseppe Scarnati and Silvio

Sala progressed the initial drawings to three dimensional 1/10th scale models and then a full-size buck. The design for the Giulia saloon was undoubtedly finalised when it was depicted in Nardiello's somewhat stylised renderings of 1962, and the basic shape of the model is credited to Scarnati: he it was who designed the inside of the car and also had some input into the Giulia Sprint GT. After rigorous testing, which included subjecting the car to colossal extremes of temperature, the Giulia T.I. went into production at the new plant at Arese on the northern fringe of Milan, while coincidentally, construction of the test facility at Balocco between Milan and Turin was well under way.

Whereas it was possible to discern a stylistic similarity between the Freccia d'Oro and 1900, thence to the Giulietta Berlina and thence to the 2000 and 2600, the Giulia looked like no other previous Alfa saloon. Indeed, there were no precedents in any European saloon car, and while the Giulietta Berlina's curvaceous type of shape was widespread throughout the industry, the Giulia T.I.'s lines were actually quite

On display in the museum at Arese - and looking rather lost in its huge case - is this pre-production clay model of the Giulia saloon, likely to have been done by model makers Giuseppe Scarnati or Silvio Sala, who progressed the initial drawings to 1/10th scale models and on to a full-size buck.

While the production lines were established at Arese, the technical drawing office was at Portello, seen here as a hive of activity sometime during the late 1960s.

exclusive. The design was finished by April 1961, complete with the characteristic scallops along the car's waistline and roof edges. Its steeply raked windscreen, wraparound rear window and Kamm-tail were the product of wind tunnel testing, the latter feature as tried and tested on the race circuit on the contemporary *Coda Tronca* version of the Giulietta SZ and the Giulia TZ of 1963. The superior aerodynamics of the Giulia saloon were featured in early sales catalogues, and its Cd figure of 0.33 is as good as, if not better than, a lot of cars made today. With good reason, the slogan used to promote it was 'The Car Designed by the Wind'.

It's difficult not to have an opinion about the shape of the Giulia saloon. Some call it boxy, drawing comparisons with the later Fiat 125 and its Lada derivative. But to dismiss it thus is to overlook a host of subtle curves, undulations and fluted lines, which combine to make it actually far more visually fascinating even than its sporting siblings. These characteristic in-

By contrast, the styling department - also at Portello - appears much more relaxed. Whilst there's no coherent rendering of a production Berlina to be seen, it's probable that they are formulating ideas for the Alfetta. One illustration seems to show a variation on the Tipo 33 sports prototype from 1969.

dentations appear to change according to how the light strikes the car, some being more obvious in raking light than others. Lest I appear a tad too defensive here, I could criticise the car's Italianate driving position, caused by the relationship of steering wheel to pedals and seat, that made a compromised driving position inevitable. On the other hand, the seats are very

The remarkable thing about the Giulia saloon's styling is the variety of scallops, impressions and embossments that embellish the contours of its bodyshell.

The subtle flutings that characterise the flanks of the Giulia saloon tend to appear different, according to how the light strikes them. One thing's for sure, it's never boring.

comfortable in that you sink into them, unlike those in my current 155 which are much firmer and don't 'give' in the same way. The Giulia's projecting floor-mounted gearlever was ideally placed for precise and satisfying shifts, and it was easier to get in and out of than the old Giulietta. I should simply sum up by saying that it's rare to find a sporting specification allied to an interesting body style in such a practical package. The Giulia could carry five people and their luggage. The spare wheel lived in a well in the right-hand boot floor, and the jack was originally attached to the right-hand inner front wing by a big wing nut. Subsequently, it relocated to the inner bulkhead of the boot.

What of that sporting specification, then? Notice the first model to be available was a T.I., with no regular Normale on offer. Well, the T.I. power unit was the 1570cc version of the all-alloy, four-cylinder twin-cam, driving the rear wheels through a five-speed gearbox. This was operated by a column shift, and there was a bench front seat, fancy two-spoke steering wheel and linear speedo. Brake and clutch pedals were floor-hinged, while accelerator was pendant. Illumination was good, with a set of headlights flanking a pair of main beam lights set in the grille. The T.I. had a side indicator with

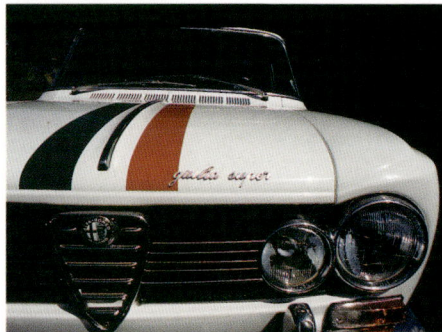

They called it 'The car designed by the Wind', so good were its aerodynamics. A drag coefficient of cd 0.33 is good, even by today's standards.

chrome embellishment over the front wheelarches. Early models had a single Solex PAIA-7 carb, and were fitted with drum brakes - three-shoes at the front - but these soon gave way to an all disc set-up. The finned alloy sump was shaped so as to accommodate a transverse box-section bodyshell crossmember. Suspension at the front was by lower wishbone, with two separate links that allowed castor adjustment acting as upper wishbones, individual coil springs and damper units. The rear was closer to that of the Giulietta, although somewhat refined, with coil springs and dampers mounted on the live axle's trailing arms, with a pivoting T-arm attached to the body via

When the model was first introduced, much was made of the Giulia saloon's Kamm tail, which Alfa claimed wind tunnel tests had proved benefited the aerodynamics in the same way as it did the Coda Tronca version of the competition-based Giulietta SZ.

The Giulia Super boasted generous carrying capacity, although its boot was not originally equipped with such niceties as a carpet, especially not one bearing the racing stripes of the Squadra Jolly Club. The jack was a practical item that was stored against the rear bulkhead.

The altogether more rudimentary dashboard of the competition-oriented T.I. Super, with its basic instrument panel, slightly dished Bakelite three-spoke wheel and simple window furniture. The area ahead of the passenger around the glove compartment is cut away in a scoop that matches some of the car's external stylistic features.

See how the Alfa shield splits when the bonnet is opened on this late Giulia T.I. Note also the characteristic grilles in which the horizontal slats are parallel and of equal size.

Rear detailing of the Giulia T.I. featured a chrome strip that encircled the rear light clusters and numberplate.

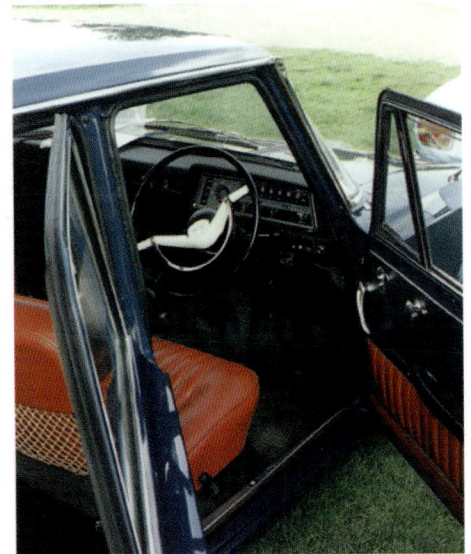

Interior of the T.I. shows off red upholstery and cream spoke bakelite steering wheel with horn ring. Its linear strip speedo is a novelty these days, while gauges and switchgear are also unique to this model. From 1962 to 1964, T.I.s had column shifts and front bench seats; the floorshift and individual seats model was produced in parallel until 1968.

Silentbloc washers providing axle location and resisting lateral movement. A canvas strap with rubber bump stops that cradled the axle and terminated upward articulation completed the picture.

By May 1964 a second version of the T.I. was available with floor change and separate front seats, and two years

This T.I. Super was being used to do the supermarché shopping run in Provence, and was acquired by its current owner, Richard Everton, in the mid-90s and brought up to concours standard by Mike Spenceley. The car is pictured at Stanford Hall on the UK AROC's National Alfa Day 1999.

The smart red upholstery of this second series Giulia T.I. belonging to Michel Rochebuet makes for a welcoming interior. It was made to special order for its first owner, and the car also has a floor-mounted shift, 'umbrella handle' handbrake lever and round dials.

One of the weight-saving areas of the T.I. Super was the front seats, which were of the bucket variety with triangular spaces at either side of the backrest. Plexiglass rear windows could also be specified for competition use.

Introduced in 1963, the T.I. Super was something of an homologation special, with just 501 cars built in its single year production run. Much of its power advantage was derived from the 112bhp Sprint Speciale engine, which used two twin-choke Webers, and the use of lightweight components. This T.I. Super is pictured in an historic touring car event at Spa Francorchamps in 1999.

later the 1600 T.I. got round instrument dials and a black bakelite steering wheel, fresh seat covers and a central armrest in the back, along with stainless steel bumpers. Production of the 1600 T.I. ended in 1968 when only one was made, so effectively it was

phased out in 1967, with 59,957 units made, excluding a few hundred in right-hand drive.

The T.I. Super

At Monza on 24th April 1963 the competition version of the Giulia saloon

was launched, designated the T.I. Super, and intended for use in the Turismo category of European and Italian races. To be homologated, it was necessary to build 1000 identical cars, and the T.I. Super was given a performance advantage by having the 112bhp engine of the Giulia Sprint Speciale fed by a pair of twin choke Weber 45 DCOE 14 carbs. Power output was some 23bhp up on the standard T.I., and this, coupled with a dry weight of 910kg as opposed to 1000kg for the regular car, enabled the T.I. Super to attain 185kph and cover a standing kilometre in just 33 seconds. The Italian racing star Geki Russo even claimed to

A Giulia T.I. Super and Mercedes-Benz 220 saloon vie for track position at the top of the rise from the ever-daunting l'Eau Rouge at Spa Francorchamps during the 1964 24-Hour touring car race.

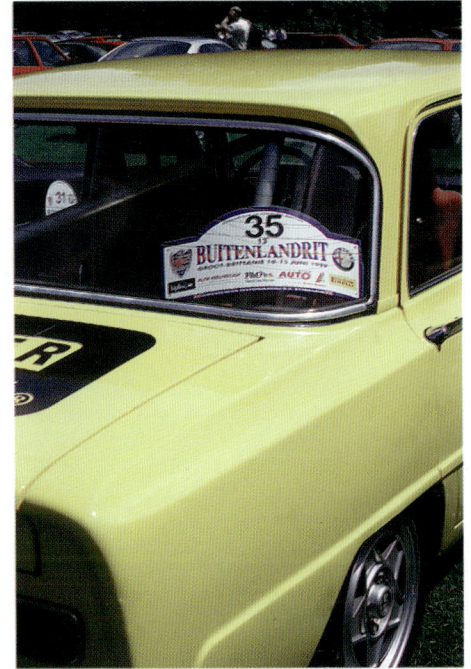

This Dutch racer was part of a large contingent from the Netherlands present at the British AROC National Alfa Day in June 1999.

have seen 200kph on the straight at Monza. With a price tag of 2.4m lira, at least twice that of the normal T.I. saloon, it had to be able to turn in something special. Private engine specialists - like Conrero, Facetti, Biscione, Baggioli, and Bosato - were employed to set up the engines for maximum output, and 140bhp was not unknown.

In the event, just 501 T.I. Supers were produced and, with only a couple of exceptions, were all finished in *biancospino* (hawthorn) white. A 1963 Argentine competition car was red, and another private car was in grey. The Milanese Jolly Club's competition cars were identified by tricolour stripes down bonnet and bootlid. According to an estimate from the Historic Giulia T.I. Super Quadrifoglio Register, only 50 cars survive today, with 31 definitely accounted for.

The T.I. Super's wheels were of light alloy by Campagnolo, with ovoid rather than round cooling holes - although in practice they didn't appear that much different to the standard pressed steel wheels. The rear windows could be in plexiglass for lightness, the bumpers were more blade-like than standard issue, and the special front seats were of a spartan bucket

type with triangular gaps at the base of the backrest where it joined the squab. Front seat belts were standard. Cabin furniture was more sporting in character than on the standard car, thanks to input from Zagato, and typified by the three-spoke steering wheel with horn button in the boss. The passenger grab handle and glovebox lid were omitted, and the instrument panel consisted of three dials, with oil and water temperature, fuel and warning lights in the centre, flanked by the rev-counter on the left and 220kph speedo to the right. There was even a basic intermittent windscreen wiper function, activated when the washer button was pressed. Like all 105-series saloons, the wipers performed their idiosyncratic arms folded motion across the screen - a style recently adopted by the Chrysler Voyager MPV, I notice.

Dunlop disc brakes were present on all four wheels, with bulkhead-mounted servo from 1964. The handbrake lever was positioned between the seats - handy for rallying and doing those handbrake turns, and the suspension was set slightly lower than normal. The main beam lights were sometimes absent, replaced by mesh shields behind the grille in order to duct cool air into the carburettors and

over the exhaust manifold. The four-leaf clover Quadrifoglio decals testifying to its sporting pedigree were prominent on the front wings.

Most of these cars were campaigned by private owners and teams, such as the Jolly Club, although the launch did coincide with establishment by Carlo Chiti and Lodovico Chizzola of the Autodelta racing team that, by 1965, had become the official works' competition arm. Alfa Chairman Giuseppe Luraghi was bent on demonstrating the company's prowess on the race circuit and formed a firm bond with Dr Chiti. Autodelta was also heavily involved with powerboat and hydroplane racing at this time, in which it was extremely successful. An unknown number of T.I. Supers were converted for competition use by Autodelta to comply with the international Appendix C regulations for modified production saloons, which, by implication, means they could have run in European and foreign touring car events. Regrettably, it's not possible to elaborate further except to say

67

Because of the rarity of the T.I. Super model, examples racing today are few and far between. However, the modifications necessary to replicate the specification, and even better it, are straightforward, and mean that regular Giulia saloons can easily emulate the real thing.

that power output was 160bhp at 7500rpm.

Winning streak

Active racers that we do know about are as follows, and the model's short competition career got off to a rather diffident start. Initially, the T.I. Super was entered for time trial events, such as the Cesana-Sestriere and Trieste-Opicina, Catania-Etna and Pontedecimo-Giovi, in the hands of Bussinello and Prinoth. The latter came first equal in the Rally of Sardinia, while Masoero/Maurin were fourth overall (behind three Ferraris) in the Gran Turismo class of the 1963 Tour de France. The T.I. Super was on its way, and although production ended in 1964, De Adamich clinched the Autumn Cup at Monza, and Baghetti took the honours in the European Cup also at Monza that year, with Munaron and Di Bono second and third overall. Munaron also shared a T.I. Super with De Adamich to finish fifth in the Spa 24-Hours in 1964, while De Adamich won the Portuguese Rally and Cavallari won in Sardinia. The Scuderia Jolly

Club car of De Adamich/Arcioni was outright winner of the annual Monza 4-Hours race on 19th March 1965, with

Pinto winning at Mugello.

Meanwhile, down under in Australia, Gardner/Bartlett shared the Alec

Dramatic contemporary photo of a Giulia Super being thrashed through the jungle during the East African Safari Rally. The bonnet has rubber fasteners, a huge pair of spotlights are mounted on the front wings, and a pair of air horns grace the bent front bumper.

Probably nowhere is the Giulia saloon as revered as much as it is in Holland, and the Dutch even have a race series exclusively for the model, in which participants belong to the Squadra Bianca - since virtually all T.I. Supers were white.

Mildren T.I. Super to win the 6-Hours of Sandown Park in 1964 and 1965, documented in a wonderful period video lent to me by Mike Spenceley. Other international successes of note were second overall in the 1965 Spa 24-Hours (Lagae/De Keyn), first overall in the 1966 International Hessen Rally (Klein), outright victory in the 1966 Corsa d'Interlagos and the 1000kms of Brazil (Gancia), and victory in the 1966 24-Hours of Interlagos (Zambello/Lolli). In the USA, future GTA racer and TransAm champion Horst Kwech raced a T.I. Super in 1965.

Surprisingly for a model that, in one form or another, remained in production for such a long time, the T.I. Super's racing history was quite brief. After that, it was the turn of the lightweight GTA coupé to come to the fore, prepared either by Autodelta or private specialists like Virgilio Conrero and Carlo Facetti. To qualify for an Autodelta GTA you had to be an estab-

Examples of the Colli-made Giulia van are as rare as the total eclipse, and this one belonged to Fernand Masoero who raced T.I. Supers in the 1960s and sold his extensive collection of Alfas in Paris in 1998. These vehicles were made as factory service vans, and tail-gate sizes varied from car to car.

lished racing driver or they wouldn't play.

During the last 10 years a club race series evolved in the Netherlands for T.I. Super lookalikes, under the banner of the Squadra Bianca. The cars comply with a fixed specification, centred on a 2.0-litre unit developing 180bhp at the flywheel, rather than the original 1600 engine. They're fitted with Cosworth racing pistons, Columbo Borriane cams with 12.5mm inlet lift

and 11.5mm exhaust lift, bigger nitrided titanium valves, DRLA 40 Dell'Orto carbs and four-into-one exhaust manifold. A ZF limited slip diff completes the drivetrain. These cars generally run with 5 degrees of negative camber at the front, and use particularly adhesive 185/55 x 14 Toyo tyres, with Intrax gas dampers and a 2 inch front anti-roll bar to keep it stiff. All-up weight is 975kg (2193lb) which takes into account the roll-cage and

perspex windows - but that still exceeds the T.I. Super's original weight by 65kg (146lb). I saw a Squadra Bianca race at Zandvoort in the early '90s and the action was tremendous to behold. I tried to buy one once, but they don't come cheap.

Coming right up to date, when GTAs are still at a premium, a truly mouth-watering collection of Alfas was auctioned in Paris in 1998. They belonged to '60s T.I. Super racer and rally expert Fernand Masoero, and included one GTA and no less than eight Berlinas of one sort or another, plus a T.I. Super and a Giulia Super Promiscua van. The signwriting on the Colli-bodied Promiscua promoted Masoero's Orange garage in Provence. One can only sigh and reflect on the possibilities that might be a reality, given that lottery win ...

Coupé derivatives

As we've seen, it was traditional for Alfa Romeo's Berlinas to have Coupés and Spiders in the same range. The Giulia saloon was no exception, except there was a glitch in the continuity of the Coupé and Spider models.

Somewhat confusingly, the 101-series Giulietta Sprint and Giulietta Spider continued in production and became identified as the Giulia Sprint and Giulia Spider when they were fitted with the single carburettor, 1570cc twin-cam from the Giulia T.I. saloon in 1962. A characteristic air filter was fitted to the taller engine so that it would fit under the little cars' bonnets, and there were numerous small differences of trim, instrumentation and upholstery. The Giulietta Sprint Speciale received the same engine to become the Giulia S.S., and was made until 1966, while the brand new Giugiaro-designed Giulia Sprint GT that nowadays we sometimes call the 'Bertone Coupé' appeared in 1963 to replace the Giulietta Sprint design. This, too, used the Giulia T.I. powertrain, augmented by a pair of Weber 40 DCOE 4 carbs; indeed, it was based on the same floorpan and suspension.

The Giulia Sprint GT and subsequent (1965) higher-performance Veloce derivative were built at Arese, while mechanicals were sourced from Portello, and the two models, saloon and coupé, went down the line in parallel and even intermingled *en route* to the paint shop. Spiders, meanwhile, were made at Pininfarina and shipped to Arese for drivetrains to be fitted. The 1600 Spider, known as the Duetto,

On the competition front, the T.I. Super's challenge faded with the rising eminence of the more specialised Lotus Cortina, and the battle was taken up by the Giulia Sprint GTA. Major savings in weight were made by panelling the car with body sections made of Peraluman.

was introduced in 1966 and successive 105-series Spiders with varying degrees of mechanical updating and aesthetic modification, remained in production until 1992.

The Sprint GT evolved into the lightweight GTA (Gran Turismo Allegerita) coupé in 1965. While the Giulia T.I. Super was saddled with the regular T.I. bodyshell, by no means a heavyweight, the GTA was furnished with Peraluman (an amalgam of zinc, titanium and aluminium) panels, which improved its power-to-weight ratio and made it far more competitive on the track. In addition, the GTA was given a twin-plug cylinder head, special conrods and larger elektron sump. It was an instant success and brought innumerable race, rally and hillclimb wins and Touring Car Championship victories for the marque over the following decade, as well as spawning a fascinating succession of ever more powerful and increasingly spectacular looking versions. These included the supercharged GT SA (*Sovralimentazione*), GTA 1300 Junior, and the wide-arch GT Am (loosely based on the US specification Spica fuel-injected 1750 GT Veloce - hence the ambiguous 'Am' suffix). These were out-and-out racing

The Junior Zagato was the last derivative of the Giulia coupé, and was made in 1300 and 1600 forms between 1970 and 1975. Just 1108 cars were made with the 1300 engine - like this one with the petrol filler on the right-hand side - and 403 with the 1600 unit. The 1600 Junior Z model is also slightly longer.

cars built by Autodelta and based on production 1750 GTVs, and the majority had steel bodies with fibreglass wheelarches, although there was no consistency about body panels - some

private builders used all plastic panels. Some were created from GTAs that had already seen plenty of action. Regular GTAs, if that's the right word, are recognisable by their ultra plain mesh grille, step-front bonnet, 6in Campagnolo wheels and the green cloverleaf on a white triangle on the car's flanks. If you've managed to get this far, you may be interested to know that one of my very first books on the subject was published by Veloce (*Giulia Coupé GT and GTA*), which is still in print, now as a paperback.

My favourite

Back in the world of Giulia saloons, an entry level Berlina was introduced on 11th May 1964. This was the Giulia 1300, powered by the 78bhp 1290cc version of the twin-cam engine, lacking a brake servo and with different gear ratios in its four-speed 'box and a single downdraught Solex carburettor. It differed visually from the T.I. in having only a single pair of headlights and seven stainless steel bars across the grille, and there weren't any overriders on the bumpers. This model remained in production until 1971, with dual-circuit brakes and a top-hinged pedal box only fitted in its

The entry-level Giulia 1300 was introduced in spring 1964, and came with a less comprehensive specification, lacking a brake servo and fifth gear. The inner pair of headlights - main beam - were absent, and, of course, the single pair took care of both functions. The side grilles in the radiator had three equally spaced prominent slats.

penultimate year of manufacture.

In 1965 was introduced a higher performance version of the Giulia 1300 saloon, identified as the T.I., which came with a five-speed gearbox and shorter 8:41 diff. The 1290cc twin-cam unit had a 9:1 compression ratio and was fed by a single downdraught Solex C 32 PAIA 7 carburettor, developing 82bhp at 6000rpm. A brake servo was fitted in 1967, along with round instrumentation, wood-effect dashboard trim and leatherette seat upholstery. There was a locking glovebox and parcel shelf on the passenger side. The three-spoke dished bakelite steering wheel had three chromed horn arms. Externally, the grille switched to three horizontal stainless steel slats on a black mesh background. Vents at the base of the windscreen in the scuttle panel were intended to allow better airflow into the cabin, otherwise you simply opened the front quarter-light windows to their fullest extent. And the bumpers now got rubber-faced overriders, which were enlarged slightly a couple of years later.

From 1969 the wider 4.5J x 14in wheels were available on the 1300 T.I., and it was equipped with a braking effort stabiliser for the rear wheels, along with revised rear suspension bushes and engine mounts, plus the hydraulic diaphragm clutch, new synchromesh and rigid gearlever. In 1970 a dual circuit braking system and top-hinged pedals were fitted to the 1300 T.I., along with the rest of the Giulia range. I have a soft spot for this model, having owned one for five or six years from 1985. It was a South African built

The 1300 T. I. version came out in 1965, and my own 1971 car was originally made as a CKD unit in South Africa - hence its survival in first-class condition. The upgraded model had a five-speed gearbox, taller final drive, and, in 1967, a brake servo. It also got bumper overriders, three-slat side grilles with a black mesh background, and changed vents at the base of the windscreen.

car, a 1972 model imported from Zimbabwe, so at the time its rather innocuous pale green 'spring water' paint finish hadn't succumbed to the northern European climate. My first outing in the car did for a valve, and it wasn't long before I'd switched to a 1600 engine from a scrapped Sprint GTV, courtesy of Alfashop in Norwich. Acquaintanceship with erstwhile Norfolk Alfa specialist Brian Hammond subsequently led to the fitting of a 2.0-litre unit and limited slip diff from a 2000 Berlina, lowering of the suspension

and fitting of Koni dampers. This produced a really excellent car for club track days and backroads fun, although it spent a good deal of time in London as I worked on a classic car magazine at the time. It was also a wolf in sheep's clothing, going, as one incredulous hack put it, 'like sh*t off a shovel', and on any kind of road it could surprise much more overtly powerful machinery. Surprisingly, the single servo brakes (actually an anachronism due to its foreign assembly) were quite adequate for stopping it, and I have to say

Lured by the quest for greater performance, my 1300 T.I. was fitted with Alfa twin-cam engines of successively greater capacity, culminating in a 2.0-litre unit and a limited slip diff. This may seem like heresy to some, but the exercise was prompted by the untimely expiry of the original motor.

tal treatment in 1970 when the step-front was replaced by faired-over bonnet, wings and front panel. Single headlights were retained, with a single trim section each side of the shield grille. In its final 1974 guise the GT 1300 Junior (as it was known from 1972 when the GT 1600 Junior became available) acquired the twin headlights and multi-slatted grille of the 2000 GTV but retained overrider-less bumpers. The road-going version of the GTA 1300 Junior - as opposed to the more intensely developed Autodelta prepared racing cars - was made in the same Peraluman 25 lightweight panels as the 1600 GTA, but with higher rear wheelarches so the 7in Campagnolo alloys didn't foul the bodywork. While the 1600 GTA was discreet, the 1300 GTA Junior proclaimed its high-performance twin-spark potential (96bhp at 6000rpm) with a quadrifoglio insignia, white racing stripe along the flanks from door to tail, suitable boot insignia and a huge Visconti serpent logo emblazoned on the bonnet.

The Giulia Super

Both the T.I. and the T.I. Super had paved the way for the definitive Giulia saloon, introduced in 1965 and logically called the Giulia Super. This model remained in production for 11 years, which demonstrated that the recipe of 1600 engine and twin sidedraught Weber 40 DCOE carbs was pretty much spot on. Supers were all-rounders, used as taxis and as squad cars by the *Carabinieri* - as seen in that wonderful Michael Caine film *The Italian Job* . Of course, we all know that, in reality, the

that I preferred the purer uncluttered look of the single headlight car over the twin-light façade of the Super. I don't know if the gruff 2.0-litre unit suited its character - perhaps the zingier 1750 engine might have been a better match. The 1300 T.I. wasn't made after May 1972 and, in fact, righthand drive cars weren't made at Arese after 1970, so

my RHD South African car was something of an exception to the general picture. Alfa Romeos were by now being assembled at seven plants in different countries, so inconsistencies like this were almost inevitable.

The 1300 T.I.'s stablemate, the Giulia 1300 Junior, had a production life of 1966 to 1975, with revised fron-

Here's an attempt at recreating some of Alfa Romeo's tasteful publicity shots that were standard issue when the Giulia range was in production. Rumours that Mike Spenceley's Giulia Super runs on 'girl power' are not without foundation ...

This is how an authentic mid-period Giulia Super engine bay should look, with air hose crossing over the cam cover and pancake filter over to the left-hand inner wing. However, such was the degree of crossover of components as one series succeeded another that amalgams of earlier ancillaries are often found with later ones.

Mini Coopers wouldn't have stood a chance against Arese's finest!

There were cosmetic changes, including the grille in 1967 deprived of the aluminium slats, and a dished steering wheel was acquired from the Sprint GT with three aluminium spokes and bakelite rim, and co-ordinated cloth or imitation leather upholstery. The early Supers had the Visconti serpent and child motif at the base of the C-pillar, but this was omitted from later cars. Mechanical updates were limited to a hydraulic diaphragm clutch, synchromesh gearbox, a braking force stabiliser acting on the rear wheels and a rigid gearlever, which meant that you didn't need to depress it before changing into reverse.

By 1970 the Super had been given a dual circuit braking system with two servos, the pedals were pendant rather than floor-hinged and the handbrake lever moved to the transmission tunnel instead of being an umbrella handle type mounted below the dashboard by the steering column, now complimented by a dished steering wheel. In 1972 the model was renamed the Giulia Super 1.6, along with its smaller-engined sibling which became the Giulia Super 1.3. They were to all intents and purposes identical apart from respective engine sizes, but then, following a stylistic facelift in 1974 - regarded by most people nowadays as retrograde - it was known as the *Nuova Super*. More on that later, because the 1600 T.I.'s replacement, the 1600 S, was ushered in in 1968, and there

Interior of Keith Booker's late model Giulia Super features dished wood-rim steering wheel, prominent instrument binnacle, angular door furniture, centre console and handbrake located between the seats.

were negligible differences to the T.I., or the Super, for that matter. Fewer paint options were available with this model than with the Super. The 1600 S initially had the five-slat grille arrangement, subsequently reduced to three slats, with small indicators on the front wings, serpent logo on the C-pillar and cabin vents in the front scuttle, and lacked a rear anti-roll bar. In 1970 it received the same dual circuit brakes and pendant pedals as the rest of the saloon range, but that was the last year it was produced. Only two were ever made in right-hand drive.

The Giulia 1300 Super was introduced in 1970, succeeded by the renamed Giulia Super 1.3 in 1972 and Nuova Super 1.3 a couple of years later. The 1300 Super was given the 89bhp engine from the GT Junior, with a pair of twin-choke sidedraught carbs, driving through the five-speed gearbox, and contemporary updates like pendant pedals and twin brake servos. Trim and upholstery levels were upgraded to Giulia Super spec as well, although the single headlights were retained with three-slat grille. As I've said, the Super 1.3 of 1972 was to all intents and purposes the same car as the Super 1.6.

Both engine sizes of the Nuova Super were equipped with twin headlights set in a matt-black grille incorporating single slats each side of a plumper chromed Alfa shield, as fitted to the contemporary 2000 Berlina. It wasn't just the façade that altered the character of the car, as the bonnet and bootlid were flattened off, presumably to give a more up-to-date look. Today, the Nuova Super looks a trifle incongruous compared with its predecessors, which all had the trademark indentations in these pressings. The most significant side issue here is that the Nuova Super was also offered with a 1760cc diesel power unit. This was the 55bhp, four-cylinder Perkins engine previously used in the Romeo F12 van, and its installation in the Nuova Super was mercifully accompanied by some five kilos of sound-deadening material. The speedo and rev-counter only went up to 160kph and 5000rpm so as

Right - The Nuova Super was given a matching pair of 136mm diameter headlights, bigger Alfa shield and single-slat side-grilles, whilst the bonnet matched the bootlid with its embossed contours now ironed out. A further 6582 units were made fitted with Perkins diesel engines.

Bottom right - Some of the hues in the 105-series colour range are very attractive, like this Pine Green Super with lowered suspension and period Minilite wheels, pictured during a club meeting at Assen race circuit in 1996.

Below: In its final form, the Giulia Super was known as the Nuova Super, and launched at Fonteblanda near Grosseto in June 1974. It was available in 1600 and 1300 engine sizes with identical bodyshells and, as can be seen here, the bootlid was flattened off in a bid to modernise the shape.

From 1974 to 1976, the long-running Super was available in this metallic olive green colour, and with yet another variation on the grille-slat arrangement. This configuration was also used on the Giulia 1600 S that appeared in 1968, and had dual circuit brakes and top hinged pedals. Keith Booker's Super also shows the switch from hubcaps to central dust caps.

Dashboard of the 1750 Berlina shows a Bakelite steering wheel - a smart wood-rim version was optional - centre console with switches and gauges angled towards the driver, and the twin binnacles of the rev-counter and speedo atop the dash-board. This car has some additional switchgear fitted on the right, but that's how it was done then if you had auxiliary lights, or whatever.

Study this picture and ponder your misspent youth! Was this a new dance craze? An excursion into psychedelia, maybe? No, our zany young couple is probably just amazed after a run in the fab new 1750 Berlina. Although it was a 105-series car, the 1750 Berlina was never known as a Giulia, which helps to not confuse it with the slightly smaller T.I. and Super models.

not to disappoint the press-on motor-ist. For some reason the diesel model was available in four colours only - white, ivory, Dutch blue or pine green.

Grown-up Berlinas

One of the factors that accentuates the Nuova's inconsistency with the rest of the Giulia saloons is the prior existence of the flutings-free, slab-sided 1750 Berlina and subsequent 2000 Berlina. The former was launched in January 1968 at Vietri sul Mare and at the following year's Turin Show. The 1750 Berlina had a longer wheelbase than the Giulia Super (2570mm to 2510mm) although track was the same - making it appear long and narrow - and it was also 40kg heavier. It was powered by the 118bhp, 1779cc version of the all-alloy, four-cylinder twin-cam, which was close enough in capacity to arouse promotions-worthy connotations with the legendary 6C 1750 Gran Sport Zagato Spider of the 1930s. Indeed, the Berlina's sisters, the 1750 Sprint GTV and 1750 Spider Veloce, were sufficiently overt in appearance to justify such comparisons, but the Berlina was a rather plainer proposition. Amazingly, it was styled by Bertone, but calculated to utilise as much componentry from the corporate

The 105-series saloon was given a facelift by Bertone and unveiled at Vietri sul Mare in January 1968. It was powered by the 1779cc version of the all-aluminium twin-cam, renowned for its sparkling performance and willing-ness to rev. The following year it was equipped with dual circuit brakes, top-hinged pedals and ram-air filter.

parts bins as possible.

Although offered with dual-circuit brakes, hydraulic clutch, quartz-io-dine headlights and pendant pedals from the outset, the 1750's unembellished bodyshell lacked the impudent charm of the smaller-engined Giulia saloons; it was a handsome enough car, but there was nothing to

make you fall in love with it, and I say this having owned one for a couple of years. It was my first Alfa, and served very well at a time when I worked in London in the mid-1970s and, for rea-sons we won't go into here, had a social life in Lancashire and Scotland.

The charisma of the plain-Jane 1750 Berlina lay primarily in its effort-

Posed like the figures in a Renaissance painting, the Arese workforce appears strangely static during production of a line of 1750 Berlina shells. Over to the left, a run of step-front Giulia 1300 Junior coupés (contemporary with the 1750 Berlina and twin-headlight 1750 GT Veloce) awaits attention.

79

less performance, which enabled it to shine at traffic lights as well as cruise the main roads at a steady 100mph. Secondly, handling around the back roads could be quite exhilarating, and thirdly, its very inconspicuousness gave it a certain anonymity when travelling quickly on motorways. It was an ample five-seater, with cloth or leatherette upholstery, and the boot of my yellow ochre car gobbled up the entire contents of my Earls Court flat in one go when I left town. On reflection, that really amounted to little more than a stereo system, but it seemed impressive at the time.

Alfa Romeo was keen to promote the Berlina as an icon of safety; the cabin area was a central survival cell and the front panels and rear floorpan crumple zones in the event of an accident. Back in the early 1960s the pre-production Giulia saloon had been subjected to a crash test programme, which showed that the cabin held up extremely well. The extra length of the car, although only slight, was in the rear doors and front and rear overhangs, and the treatment of the C-pillar enabled them to fit a much shallower rear screen. The model was in production until 1970 although, typically, there were overlaps. In this case, the model was equipped with Spica fuel injection for the emissions-conscious US market, and that was available until 1971. The dashboard featured two binnacles ahead of the driver containing the principal dials, with auxiliaries mounted in a central console angled towards the driver.

The 1750 Berlina was seen less

frequently in competition than was the Giulia T.I. Super, partly because it was a slightly larger and more opulent car, and, in any case, the ubiquitous GTA was doing the business at the time. However, in the 1968 Alpine Rally, the 1750 Berlina of Bianchi/Jacquemin won the production car class, and in July 1968, the four 1750 Berlinas entered scooped the first four places in the Group 1 production category at the Spa Francorchamps 24-Hours. The following year, the Group 1 Alfa Romeo Benelux 1750 Berlina of Wolleck/Engemann placed 13th in the same event.

Perhaps unwittingly, perhaps by design, the 1750 Berlina was used as the mule for the GT Am that entered the Touring Car fray with such impact

in 1970. Also featuring in the 1969 Spa marathon was a 1750 Berlina with a 1988cc twin spark engine and specially adapted GTA head, prepared by the Belgian engine builder Pierson and driven by Demol/Petre. It's certain that Autodelta was already on the case, but the Belgian Berlina seems to have been the first documented adaptation of the power unit that would feature so prominently in the bulbous-winged racing

1750 and 2000 GT Am coupés. Evidently they were on to something, as it endured for 18 hours before the head gasket gave up.

Berlina torque

In 1971 the 2000 Berlina was launched to replace the 1750 model; apart from the larger engine, the only external differences centred on a fresh frontal aspect, with the broader Alfa shield grille mentioned earlier, and, like the Nuova Super, hub caps were abandoned in favour of exposed wheel nuts and dust covers with the Alfa motif at the centre. Sidelights were larger at the front, and the whole light cluster at the rear was revised and enlarged. It's entirely probable that the capacity increase was prompted by the huge success of Autodelta's 1750 GT Ams in European Touring Car racing (Toine Hezemans was European Champion in 1970), which ran with 2.0-litre engines. Not only the Berlina,

Here's a good cross-section of Giulia saloons pictured at a club gathering. From right to left, they are: 1300 T.I., Super, Nuova Super, 1600 S and Super, etc.

but also the 2000 GTV and 2000 Spider Veloce were fitted with the larger bore 1962cc twin-cam unit. The Berlina could be specified with a limited slip differential, and a pair of Solex C 40 DDH or Dell'Orto DHLA 40 sidedraught carbs. Because of their relatively light weight all the 105-series saloons were quick off the mark, with around 10 seconds for the 0-60mph dash about par for all three models - 1600, 1750 and 2000, on the basis that you don't burn the clutch out like they do calculating road test statistics for motoring magazines. The 1300 took more like 13 seconds to get there. Top whack was around 110, 115 and 120mph for each of them, with just under 100mph for the 1300. As with the 1750 Berlina, the 2.0-litre car utilised the same live rear axle and suspension layout as the Giulia Super, with very slight alterations to the geometry.

Front seat headrests and a heated rear window were also available in the 2000 Berlina, features taken for granted as standard for years now. In 1975 the major mechanical update was the fitting of transistorised electronic ignition, which improved cold starting - no need to fiddle any more with the characteristic choke and hand throttle levers that lurked below the ignition switch. The interior trim level was further improved over that of the 1750 model, and a classic wood-rim wheel fitted, while instrumentation was incorporated into a more coherent scheme within the dashboard, although the centre console remained. Rear head restraints were fitted, and the cabin interior was quite palatial. An electric screen washer system replaced the earlier version operated with a foot pump, and a new gearknob and steering wheel boss were also fitted.

The model went out of production in 1976, and it's worth mentioning that the Spica fuel injected cars exported to the States were known as 115-series cars as opposed to 105-series, and that between 1972 and 1975, automatic transmission was available. Production figures for the 105-series saloons were 542,085 for the Giulia in all its forms. Most numerous was the 1300 Super, at 195,031 units. Then there were 101,883 units of the 1750 Berlina, followed by 86,245 units of the 2000 Berlina.

To my mind, the Giulia Super epitomises Alfa's classic saloons, especially from the 1960s, as I feel the company lost its way a bit with the 1750 Berlina, and the Alfetta that followed was even more fashionably slab-sided and angular in design. Going further back in time, the Giulietta Berlina isn't as roomy or as powerful as the Giulia - added to which it just isn't as plentiful any more - so rarity becomes an issue. On the other hand, I have proved that you can run a classic Giulia saloon quite happily on a day-to-day basis, and spares and servicing were never a problem.

Alfa Romeo Berlinas

7

AN ALFA FOR THE PEOPLE

If Alfa Romeos were traditionally owned by comfortably-off motor sport enthusiasts - a sweeping generalisation, granted, but not without some justification - then things were about to change. If any Alfa saloon captured the imagination of the general public it was the Alfasud, whose spiritual successor, the 33, was produced in the greatest numbers of any Alfa Romeo to date.

By 1970, as part of the grand plan, the Alfasud plant was under construction at Pomigliano d'Arco. Alfa's aeronautical engine plant had been at Pomigliano since the 1930s, but the prospect of a cheaper and more plentiful workforce, aided and abetted by governmental influence, led Alfa Romeo to choose the southern site - from which the 'Sud took its name - at which to build it. And incidentally, due to the factory's proximity to Naples, the word Milano no longer featured in the Alfa logo.

The project was overseen by one-time Cisitalia engineer Dr Rudolf Hruska, an Austrian who had also worked with Professor Ferdinand Porsche on the original Volkswagen. Hruska had also been closely involved with the launch of the Giulietta in 1954, had then resigned and gone to work for Fiat before getting the call

Straight off the production line, a number of Alfasuds are put to the test on the track at the Pomigliano d'Arco plant. This consisted of a huge circuit with banked bends, interspersed with a mixture of smooth tarmac, interim concrete and cobbled pavé, and sundry hazards like a humpback bridge.

The Alfasud arrived a little late to catch the kudos of hippie chic (which belonged to the Mini, or maybe the VW Beetle), but its modernity meant it was a genuine hit with the early '70s generation, as portrayed in this Alfa publicity shot around the time of its launch in 1971.

from Dr Luraghi to come back and direct the engineering aspect of the Alfasud. His right-hand man was Ing Domenico Chirico.

New plant installed at Pomigliano to build Alfa's first utility car included metal presses and a certain amount of robotisation, which was in-line with the times but somewhat at odds with the original motive for locating the plant near the labour-rich city of Naples. By the time the 33 came out, the Pomigliano lines were fully robotised. Even the painting process was automated for the 33, as the shells were drawn along an aerial track like trolley buses. The pigment was applied by way of rollers from above and rotating cups from the sides. After the vehicles had been assembled they were put through their paces on the Pomigliano proving ground, which included a high-speed circuit with banked bends, a variety of different road surfaces and a humpback bridge.

But we're jumping ahead here. The concept of the Alfasud was very much in the spirit of the times: a compact front-wheel drive, four-door, four-seater car with a flat-four engine. Citroën was simultaneously working on its 1015cc flat-four-engined, award-winning GS model - a roomier car than the 'Sud, boasting hydropneumatic suspension, and enveloped in an attractive fastback body shape. The Alfasud's shell was drawn by Giorgetto Giugiaro and was a marriage of form and function, in which sense it was a masterpiece.

After much testing in a variety of environments - including the north African desert and the arctic regions of Sweden - the Alfasud was unveiled in November 1971 at the Turin Show. Deliveries began in June 1972, and it's worth bearing in mind that the 105-series tin-tops were still going strong, in the guise of the 2000 Berlina and GTV at this stage, so the 'Sud was a very different proposition indeed.

The Alfasud's 1186cc, flat-four 'Boxer' power unit featured a camshaft at the end of each bank of cylinders, and valve gear driven by toothed belts, with a single Solex C32 DISA/21 carburettor feeding long intake manifolds. The Alfasud's engine bay was characterised by a bulkhead that separated the engine from the battery, heater, fuse box and fluid reservoir. The power unit was located ahead of the suspension turrets, with the radiator's reservoir mounted above the left-hand inner wheelarch. Having the weight of the flat-four engine distributed low down in the car made for a lower centre

Anatomy of the Alfasud's flat-four engine, showing the location of camshafts at the outer extremities, horizontal pistons, frontal drivebelts and central crank, with carbs and air box overhead.

of gravity than could be obtained with a regular in-line engine. Although its forward location inevitably meant that the car had a frontal weight bias, it was nonetheless a compact and efficient design.

Because the engine was located so far forward in the monocoque shell, the space for occupants in the cabin could be relatively greater. Seats were state-of-the-art for a mass-produced car, with headrests and reclining facility in the front. The original dashboard layout and controls of the Alfasud were a model of simplicity, and, apart from the choke and windscreen washer, all other controls were operated by two stalks on the steering column: lights and indicators to the left and heater, wipers and horn to the right. As we'd come to expect, the driving position suited shorter drivers with longer arms, and on right-hand drive cars, the pedals were offset to the left to give clearance to the front wheelarch.

Suspension consisted of conventional MacPherson struts, coil springs and dampers, plus anti-roll bar, with inboard disc brakes at the front. At the rear was a lightweight rigid axle with semi-trailing arms and a transverse rod, plus coil springs and dampers.

The new model was 14-inches shorter and 450lb lighter than the 2000 Berlina. Its low centre of gravity and wide track meant that roadholding was superb; the car being easy to drive quickly because of neutral handling and a good response to lift-off oversteer. It cornered as if on rails, and quickly gained the reputation of being the sort of car that you went for a drive in just for the hell of it.

Innovations swiftly introduced were a brake servo in mid-1973 and a steering damper later that year. First derivative was the *Lusso* version, known simply as the Alfasud L, launched in 1975. Apart from a few additions to the external trim, the differences hinged on the provision of ashtrays, an electric screen washer and optional rev-counter. Performance

was unchanged, but the entry-level model with a four-speed gearbox was known as the Alfasud N - for Normale.

It was only a matter of time before the suffix that had identified sporting versions of the Giulietta and Giulia Berlinas in previous decades was applied to the 'Sud, and the two-door Alfasud ti model was announced in 1974, equipped with a largely cosmetic boot-spoiler and embryonic air-dam at the front. There were now lower case initials for the *turismo internazionale* abbreviation. If the original Alfasud could be described as inadequate in any way, hindsight says that it was let down by poor top-gear acceleration due to a lack of torque at low revs. This deficiency was redressed with the 'Sud ti, and was accomplished by raising the compression ratio slightly from 8.8:1 to 9.0:1, and replacing the single-choke Solex of the standard car with a twin-choke Weber 32 DIR 41 carburettor. Although output rose only from 63bhp to 68bhp, the improvement in performance was in reality more pronounced than the figures suggest. The two-door Alfasud ti was equipped with a five-speed gearbox, and a rev-counter was provided in this model, which was a good idea as the relatively modest 1186cc engine responded best to hard work.

85

Shooting Brake

Between 1975 and 1981, an estate-car version of the 'Sud was available. Known as the Giardinetta, it was undeniably a practical vehicle with a long, flat cargo deck when the rear seat was folded down, but while its mechanical underpinnings were clearly those of the Alfasud, the Giardinetta was not an aesthetic success. Apart from head-on, it contrived to look much like any other small estate. And, quite naturally, when fully laden, much of the liveliness of the car disappeared. Engine capacity increased in 1978 to 1286cc and soon after to 1351cc which helped performance under load. If you were into Alfasuds and needed to carry cargo, the Giardinetta was for you, since at this point in its history the regular car's boot was hinged just below the rear window and had yet to blossom into a hatchback. It also lacked a strut to prop the lid up, so that had to lie back against the rear window, which implied trouble on a windy day. In that respect alone, the Alfasud was showing its age, compared with that paragon of hatchbacks launched in 1974, the VW Golf.

It's also worth taking stock of where Alfa Romeo was at on the bigger world stage around this time. It was still active in powerboat racing, and continued to produce its Tipo A12 light-duty truck and, in 1972, the four-wheel drive A38 came out, which was

something like a Steyr-Puch Haflinger or Mercedes Unimog. The T33 sports prototypes and 1750 GT Ams were doing well in their respective arenas of competition, and the Montreal was its representative in the classy exotic segment, in the early '70s, at least.

The 'Sud was a sound basis from which to explore different market segments, and Alfa Romeo launched other versions accordingly. The 1490cc version of the Alfasud ti - the 1.5 - was introduced in 1978, fitted with the same 85bhp engine as the Alfasud Sprint coupé, and this model received progressive styling modifications and capacity revisions up to 1981, when it was rated at 95bhp. The four-door 5M of 1976 was fitted with the ti's five-speed gearbox, and had the higher level of trim found on its predecessor, the L model, including new seats, cloth trim on the door panels and carpeting on the floor. Instrumentation consisted of a pair of pale blue dials and a centre console was installed. Meanwhile, the basic N version retained its austere

'When I said bring the flares I meant just the cartridges!' Yes, you could shoot off to the country loaded up to the gunnels in the versatile Alfasud Giardinetta. As well as being completely practical with a full height rear door and foldaway rear seat, it embodied all the mechanical innovations of the Alfasud saloon. A total of 5899 units were built between 1975 and 1981, when it was made redundant by the introduction of the hatchback saloon.

The Alfasuds run in Classes E and F in the UK's AROC Championship series, and also, in more highly modified form, in classes A to D. Here, in April 1999, Class F midfield runner Steve Fletcher is pictured at Snetterton in his four-door 'Sud ti - a rare bird for a race car, as generally two-door models are campaigned for reasons of lightness.

interior. In December 1977 the two- and four-door 1.3 Super version was introduced, with engine capacity increased to 1286cc. The Super got twin carbs in 1979 and taller gear ratios.

Following criticism about corrosion, more rustproofing was applied to the bodyshell from 1977, while larger bumpers were fitted, with sprung polypropylene sections, and window trims were in stainless steel. Front grille, bonnet air intakes and C-pillar vents were now matt black. Also in 1977 the second series Alfasud ti was given the 76bhp, 1286cc engine that powered its recently-introduced sibling, the Alfasud Sprint coupé, and it became known as the Alfasud ti 1.3. This longer-stroke engine, combined with the five-speed gearbox and breakerless electronic ignition, gave a big improvement in acceleration right the way through the rev range, as well as making for better economy. It was a more relaxed car to drive as it was no longer absolutely necessary to make such frantic use of the gear stick. Only a year later, capacity was upped to 1351cc by raising the stroke, which

lifted output to 79bhp, and when the Sprint 1.3 Veloce engine was installed in 1979, the 'Sud ti was rated at 86bhp. The specification of the 1.3 Alfasud ti's flat-four engine included twin downdraught carbs, concave piston tops and single row of valves in the roof of the combustion chamber, and single camshafts located on each side of the power unit.

The multi-finned Campagnolo alloy wheels (a real hassle to clean) were fitted to enliven the top-of-the-range Alfasuds, and a revised boot spoiler added to the enhanced sporting image. But these trappings weren't quite good enough to keep the model above criticism. Whereas seating and controls were originally considered acceptable, and any deficiencies counterbalanced by excellent handling and performance, by 1978 certain aspects of the interior were criticised for being cheap and flimsy, and fittings crude and poorly finished.

By the end of the decade, successive price increases gradually whittled away the original Alfasud's clear price advantage over anything offering com-

parable performance and practicality, such as the Citroën GS and VW Golf, to the extent that buying one was a matter for the heart rather than the head.

Waiting for Junior

Things began to look up with the new decade. In 1980 Alfa Romeo fitted the 1.3 Super with a twin-choke carb, lifting power to 62bhp, and serious facelifts were being administered throughout the entire model ranges. The Alfasud was no exception, getting matt-black wraparound plastic bumpers and revised tail-light clusters that now overlapped the bootlid, a boot spoiler and a plastic air dam that formed the front valance. This third series also got Alfa Romeo badges mounted on the C-pillar. The external facelift also carried with it an updating of the Alfasud dashboard. All the controls were basically the same as before, but the overall design of the instrument panel gave it a more contemporary look. By 1980, car makers had conceded that owners needed to carry more than just the shopping, and the rear seat was adapted so that longer items such as

87

From 1977 the Alfasud ti inherited the 1286cc powertrain from its Sprint sibling, and the following year it was available with the 1.5 unit. The car pictured is a 1980 ti model, with twin headlights and boot spoiler fitted.

Above right - By 1981 the Alfasud was beginning to lag behind competitors like the VW Golf and Citroën GS, and introduction of a hatchback was long overdue. When it arrived, it transformed the 'Sud into a much more realistic proposition for motorists shopping for a hatchback model.

Below - Clearly this is not a driving school, otherwise this haphazard parking wouldn't have been tolerated. Seriously, though, this is the Alfasud Junior, an attempt to move some metal in 1982 by evoking memories of the similarly-named Giulia 1300 and 1600 GT models from the '60s and '70s with this special edition entry level car.

skis could be carried lengthways within the car. The following year, the issue was properly addressed by the introduction of a true hatchback. In addition, the rear window could be kept clean with the new wash-wipe system. The car was now vastly more practical and had clawed back some of the ground lost to market rivals.

Nevertheless, it was still necessary to slap on special edition labels in order to attract buyers, and in 1981 the Valentino-badged 'Sud, named after a fashion designer, appeared briefly. Rather more apt was the limited edition Alfasud Junior which came out in 1982, following the example of the 105-series Giulia coupé range that featured the 1600 and 1300 GT Junior and the 1300 GTA Junior during the early-'70s. Just 6000 of the Alfasud Junior were produced and it was priced at 650,000 lira below the regular 1.2 Super version. The Junior incorpo-

rated more rustproofing, and was identified by additional stripes and badges that were intended to impress the boy racer. Sorry, that should read 'younger driver'. Actually, it looked rather good, although I never liked those black plastic 116-series Giulietta-style wheel trims that characterised all 'Suds of that era.

Left - Some say that the original Alfasud concept was never bettered - small, compact, high-revving and excellent driveability - but there's an argument with aesthetic consideration on its side that says the 1982 ti Green Cloverleaf was the cream of the crop. Powered by the twin-carburettor 1.5-litre boxer engine, it could deliver performance and handling that was second to none. Shame about the corrosion ...

The coupé version of the 'Sud was the Sprint, in production from 1976 to 1988, seen here in its latter 105bhp Quadrifoglio Verde guise, post-1983. Designed by Giugiaro, it could pass for a scaled-down Alfetta GTV, with which it was often confused by the uninitiated, and was a neat design that encompassed all the virtues of the regular Alfasud saloon. Its tailgate was more practical than that of its GTV stablemate as it was supported on twin struts rather than one central one.

A pair of nimble 'Sud tis driven by Charles Hill and Colin Wing overhaul Paul Jaggard's 2000 Berlina at Cadwell Park during a round of the Alfa Romeo Owners' Club Championship in 1988. When racing an ostensibly bigger-engined car - like a 2.5-litre GTV6 - it was somewhat galling to be passed by an Alfasud with only 1.5-litres, but that's how effective they could be in race trim.

A hectic moment at Snetterton's Russell Corner as Ian Brookfield's 'Sud ti heads a pack consisting of a 75, Sprint and 33 around the clenchingly tight right-left complex. This was a round of the 1997 Auto Italia Championship, which Ian won that year.

This aesthetic shortcoming was addressed satisfactorily in 1982 by the 'telephone dial' alloys (shod with TRX 190/55 HR 340 or P6 185/60 HR14 tyres) that were fitted on the three-door Alfasud ti Green Cloverleaf. This shortlived model was a three-door hatchback, powered by a twin-carburettor version of the flat-four, 1500cc engine first used in the Alfasud Sprint coupé in 1979. Power went up to 105bhp with the Green Cloverleaf ti, thanks to polished ports, new cams and revised carb settings. To my mind, this is the definitive Alfasud, as it looks the part and goes very quickly too. Purists will disagree, believing that the original version was never improved upon, but you have to admit that it looks a tad weedy next to the twin-headlight ti model, and there's no sub-stitute for the increased capacity of the 1.5 that brings an instant gain in performance.

In May 1982 the four-door 1.3 Super was dropped in favour of the five-door hatch, and the 1.3 Super was also available in three-door hatch format. Trim levels were improved with wash-wipe facility, and the SC (Super Comfort - in an Alfasud?) version gained in the interior furnishings department. The Gold Cloverleaf model was awarded electric windows, armrests in imitation wood, different seat fabric and a new steering wheel. The 1.3 SC in three- and five-door forms was phased out in September 1984, while the 1.5 Super, which had been available in three- four- and five-door format, demised in 1983. Total Alfasud production, including the Giardinetta but excluding the Sprint models, was 906,803 units.

The Alfasud Sprint models endured until 1987, and - more than ten years after - the 'Suds and just the occasional Sprint were still being raced in the Alfa Romeo Owners' Club race series. What's more, they were still winning. When I raced my GTV6 in the late 1980s, I was, let's be honest here, a back marker, and in my class there were usually a number of 'Suds - with a whole litre's deficit - ahead of me. The Class A cars on slicks were lapping me, which was, at first, an appalling, if not humiliating experience. It's much the same story at the beginning of the new millennium, with Alfasuds and 33s generally quicker than the rear-wheel drive models.

Alfa Romeo
Berlinas

8

MORE
METAL FOR
THE MONEY

Pictured here at Derwentwater in the English Lake District, the Alfetta was at first powered by just the 1.8-litre engine, which, although it was a sweet, free-revving unit, lacked the gruff torqueyness of the larger 2.0-litre unit.

Rather surprisingly, the new Alfetta saloon was announced in 1971, the same year that the 105-series 2000 Berlina came out. Although powered by the 1.8-litre version of the all-alloy twin-cam, there appeared to be no reason why the Alfetta could not supersede the 105-series at a stroke. As it was, the old '60s design lasted until 1976. In practice, the Alfetta wasn't actually available until May 1972 because the company gave precedence to the launch of the Alfasud.

While the Giulia was a logical name to choose to follow on from the Giulietta, the company chose to go back to its halcyon days to find a name for the new model. After all, when you've had a winning formula, there's no harm in capitalising on it, so the Alfetta name that had graced the (virtually) invincible Tipo 158 and 159 Grand Prix winners of the early '50s was revived for the new saloon car. Principal justifica-

The new Alfetta was named after the invincible Grand Prix cars from the postwar era, a title justified by the transaxle driveline and De Dion rear suspension.

An early exponent of the Alfetta's abilities in competition in the UK was Jon Dooley, seen here at speed on a stage of the Avon/Motor Tour of Britain in August 1975.

What's a respray between friends? Jon Dooley in the same car, now in blue and orange, takes to the kerbs at Thruxton's chicane during a round of the Radio 1 Championship in 1975.

tion for this was that both vehicles utilised the transaxle gearbox principal and de Dion rear suspension, but there, predictably, the similarity ended. It was, however, the first time a production Alfa had been fitted with de Dion suspension, consisting of a Watts linkage that pivoted around the centre of the axle to help ensure that the rear wheels remained as upright as possible under cornering. Coil springs and dampers were mounted separately from one another, and the rear disc brakes were located inboard alongside the alloy transmission casing. Housed in the transaxle assembly were the differential, transmission and a flywheel with a self-adjusting hydraulic clutch. A limited-slip diff was optional at first and became standard later. The transaxle was effectively positioned underneath the back seat, and this configuration endowed the car with a weight distribution that was evenly split 50/50 front to rear, which meant that it felt a bit more 'planted' on the road than did its predecessors. At the front was another novelty; long torsion bars replaced the coil springs normally fitted. There were lower wishbones, and an ovoid upper arm through which passed the damper unit, plus castor rods and an anti-roll bar. Disc brakes were outboard at the front, and the Alfetta 1.8 was the first Alfa production car to use ZF type rack-and-pinion steering.

Trieste launch

Like the Giulia range, the 116-series Alfetta was built at Arese. Launched at Trieste in 1972, the Alfetta was a compact design: its three-box shape an uncluttered collection of largely flat planes and crisp angles in keeping with the styling trends of the time. The wheelarches were fully round, in contrast to those at the rear of the 105-series cars which were finished in a sort of spat-like way. Overall, proportions weren't dissimilar to those of the 105-series 2000 Berlina, but the Alfetta lacked the charm that, today at least, we perceive in the Giulia Super. The façade of the original model featured twin headlights set in a black grille, with a trio of chrome slats each side of the narrow Alfa shield. Chrome bumpers were fitted with rubber overriders, and there were chrome surrounds to the side windows. Giulia-style wheels with ovoid cooling holes and exposed wheelnuts were retained.

The first Alfetta saloon was powered by the familiar aluminium four-cylinder twin-cam, initially of 1.8-litres - the old 1779cc unit, popularly known as the 1750 - with a deeper, squarer profile sump. It was fed by two twin-choke sidedraught Dell'Orto carbs, and the model was delivered to the USA in 1974 fitted with Spica fuel injection.

Inside, the Alfetta's interior was more spacious than that of the 105-series models. The dashboard was logi-

The Alfetta 1.8 Berlina was by far prettiest in original configuration - with shallow chrome bumpers, twin headlights, simple three-slat side-grilles and modest Alfa shield - before the dreaded plastic and rubber protection hit it.

cally laid out, although it was somewhat cluttered and lacked the style of the old Berlina's main instrument binnacle. However, the driving position of the Alfetta was still not sufficiently adaptable for the taller physique of the typical northern European driver, so that the splayed knees attitude was *de rigeur*. How often on a long journey in an Alfetta I have longed for a foam knee rest mounted on the trim panel of the driver's door. The most frustrating function of all was the gearshift, which was vague and imprecise due to the long linkage back to the transaxle at the rear. It bore no comparison with the delicate precision of the 105-series shift, although when up and running in second and third gears, all that was required was time to accomplish the throw from one ratio to another. First to second, especially, couldn't be rushed, otherwise you'd incur the dreaded graunch of reluctant meshing, and slotting into reverse in a hurry provoked a similar reaction. Some specialists have been known to improve the quality by lubricating the gearlever pivots or fitting shims in the linkage.

Giugiaro design

The Alfetta 1.8 was joined in 1974 by the suave Alfetta 1.8 GT coupé model, credited to Giugiaro but apparently spurned by him because of in-house alterations made to the design. Whatever his view, I happen to think the Alfetta GTV shape - especially in its chrome-bumper format - ranks as an all-time classic design, superior to

many more highly fêted machines - and that's in spite of the gearchange and its propensity to rust!

The Alfetta GT in 1.8 form was only in production until 1976, and was superseded by the 1.6 GT, which lasted until 1980, and the GTV 2000 that endured until 1985, complete with the rubber bumper makeover in 1983. On the competition front, the Alfetta GT was immediately pressed into service as a rally car, clad with curvaceous wheelarch extensions. Autodelta prepared a team of nine Alfetta GTs, recruiting top exponents like Andruet, Ballastrieri, 'Biche', Dini, Fagnola and Zanetti to tackle the international rally series.

Despite promising results, Alfa Romeo's perilous financial circumstances led the company to pull the plug at the end of the year - just after signing an agreement with Intersind and the Engineering Workers Federation (*Federazione Lavoratori Metalmeccanici*, or FLM) to implement measures to stave off economic disaster, making over 5000 workers redundant in the process. Anyway, some of the Alfetta GT competition cars were sold to the Jolly Club, and in private hands on the race circuit the Alfetta GTV arrived as a welcome breath of fresh air in the wake of the long-running 105-series GTA derivatives that were still astonishingly successful in the mid-1970s.

In January 1975 the Alfetta 1.6 saloon was introduced as an entry-level model, with single headlights and

a solitary chrome slat across the grille, similar to the grille treatment of the 1974 Giulia 1600 GT Junior. Economies were made in a number of ways, such as omitting the map pockets on the backs of the seats. There were no overriders, the wipers were matt black and the steering wheel rim imitation leather rather than wood. Despite its 1570cc power unit (derived from the Nuova Super), the Alfetta 1.6 gave away only 5kph to its 1750-engined sibling in top speed. It received the updating and facelifting that was subsequently applied to the rest of the range, and was in production until 1983.

First modifications to the Alfetta 1.8 came in 1975 with better carburation that improved fuel economy, albeit at the expense of some 4bhp. There was now a much larger and more prominent Alfa shield at the front, the three horizontal chrome grille slats were removed and an external rearview mirror fitted. Inside, the driver got an adjustable head restraint and electric screen washers.

The robots arrive

By 1977, when the first robots were being installed on Arese production lines, the Alfetta saloon range was expanded to include the 2000, which had the same engine as the GTV model. It was distinguished by a facelift that included revised rubber-clad bumpers, intended to satisfy safety legislation, instead of the more elegant chromed ones on the early cars. North American market cars - known as Alfetta Sport Sedans - were further burdened by mandatory heavy-duty 5mph-impact

The Alfetta 2000 was introduced in 1977, and the range was given a major makeover that included rectangular headlights, rubber-faced bumpers (that had taken on new proportions to satisfy safety legislation), grille, scuttle air vents, door handles, wheels and rear lights.

bumpers. The Alfa grille-shield was now broader, set in a matt-black aluminium grille, and rectangular headlights further compromised the model's good looks. Door handles and rear lights were changed, side indicator units moved further back, the driver's door mirror was electrically adjustable and, inside, the instrument panel and steering wheel, seat shape, upholstery and ventilation system all received a makeover. The L-model had a fully opening sunshine roof.

The following year revised gearing and new 9.5mm-lift camshafts were

The Alfetta was produced with a VM turbodiesel engine between 1979 and 1984, in two sizes - 2.0- and 2.4-litres. Designed as a marine diesel, this unit breathed via a KKK turbo that made progress less onerous than might otherwise have been the case. The bulkhead between engine bay and passenger compartment was lined each side with sound-deadening material.

fitted, which had the effect of increasing power output by 8bhp and reducing fuel consumption. Late in the year, a three-speed ZF automatic transmission was offered, which at least meant that you couldn't complain about the ponderous gearchange. The automatic models were unique in having a hydraulic self-levelling suspension facility that ran off an engine-driven pump. They also had a transmission fluid warning light and push button tester.

Between 1979 and 1983, a 2.0-litre Turbodiesel engine was available - the first in an Italian car - with a brand new TD engine of 2.4-litres fitted between 1983 and 1984. This was the unit produced by Finmeccanica subsidiary VM at Cento near Ferrara, featuring a KKK turbocharger producing 0.85 bar boost pressure at 2600rpm. Top speed was 165kph and, since the oil-burner was equipped with all top-of-the-range accessories, including extra sound-deadening, it must

Another example of an Alfetta being used in competition, here in 1988 in the hands of Mike Barfoot, rounding Druids hairpin at Brands Hatch in the tyre tracks of David North's slick-shod Alfasud ti. By contrast, the 2.0-litre Alfetta is running standard wheels.

have been intended for a market segment that liked the image but didn't need the performance.

Bumper crop

In December 1981 the Alfetta range was facelifted and all four models - 1.6, 1.8, 2.0 and 2.0 Turbodiesel - used the same shell. Surprisingly, electric windows were standard on the Turbodiesel only, whilst the 1.8 got headlight wash-wipe facility and head restraints and seatbelts were standard. In 1982, a Gold Cloverleaf model was introduced, equipped with US-spec Spica fuel injection, and fitted with four headlights and Campagnolo alloy wheels of the dished plate variety. The following year, this model switched to a more sophisticated Bosch Motronic electronic fuel injection system and a variable valve timing facility that adjusted valve timing according to engine speed.

A further revision came in April 1983 and the 1.8 model acquired polypropylene bumpers, grey mouldings along the lower sides, black plastic wheel centres, a plastic grille and black plastic air vents and panel between the rear light clusters. Externally, it was not a pretty car anymore, although the interior had matured somewhat, and featured a wood-rim steering wheel with large rotund boss, and a gearlever knob shaped like a golf club head. The switchgear was still located rather haphazardly, but the overall effect was to make the controls seem better co-ordinated. The specification now included headlight wipers with a washer system, and electric

windows and central locking became standard. The 1.8 Alfetta went out of production in 1985, outlasting its 2.0-litre sibling by a year.

Two other versions of the Alfetta GTV bear a mention here, even though they aren't, strictly speaking, saloons. One is the shortlived Autodelta-built GTV Turbodelta, made between 1979 and 1981, which used a KKK turbo to lift power to 150bhp against the standard model's 122bhp. Secondly, the GTV6 version with the 2.5-litre V6 engine was made between 1980 and 1986, heralding the dramatic switch to plastic bumpers and front air-dam that would b eused on the GTV 2000 in 1983. Although undeniably more powerful and sonorous, the GTV6 was less attractive than the GTV 2000 because of its bonnet hump, and less agile because of its greater weight up front. However, the GTV6 did the business on the race track, providing the bulk of the midfield runners in European Touring Car racing throughout much of the 1980s. The European title was snapped up by Andy Rouse in a GTV6 in 1983, and Alain Cudini was French Production Car Champion in a GTV6 the same year. Nowadays, Alfetta fanciers turn their attention to upping the performance factor of the 2.0-litre cars by fitting redundant Twin Spark engines from wrecked Alfa 75s. I like that idea.

Nice name, shame about the shape

When a name for Rudolf Hruska's Alfetta-based design for another saloon car was needed, the clock was turned back a decade or two and the Giulietta

nomenclature came up again. Whilst the earlier range had been a rather charming, soft-edged series that broke new ground in terms of styling and performance, the latter car was a belated attempt at cashing in on the '70s wedge profile, using tried and tested components from the Alfetta parts bin. Appreciation of all styling is subjective but, like Oliver Winterbottom's Lotus wedges of 1974, I wouldn't say the 1977 Giulietta was a success in that respect. I don't think you can view it from any angle and say, hand on heart, that it really works. Its curtailed boot meant it was less practical than the Alfetta, and, although I've only driven a 1.8 model, I don't believe there's anything to choose between the two in terms of handling and ride. So I'm left wondering what the point was, apart from the fact that Alfa could, and did, offer it as a 1.3 model. This, however, proved inflexible at low revs, and achieved maximum power at a screaming 6000rpm. Weighing in at 1020kg as opposed to the Alfetta 1.8's 1040kg, there may have been a slight advantage there, but nothing significant. You have to conclude, then, that a market was visualised for a squatter version of the Alfetta, with more angular styling and truncated tail. Although the suspension and running gear were identical, external features - like the lip at the trailing edge of the boot and the slim front air dam - were sufficiently suggestive of a more sporting character.

Launched in 1977, the 1.3 and 1.6 models shared the same cabin interior, featuring seats upholstered with

Rear seats of the 1982 Alfetta are hard to fault nearly 20 years on, with their practical and hard-wearing cloth upholstery, comprehensively equipped with central armrest and diagonal seatbelts with central lap belt.

Above, right - By 1982 the Alfetta 2000 was five years old, and the facelift that ushered in the Gold Cloverleaf version, like Keith Booker's car pictured here, brought with it a front air-dam and two-tone paint scheme. New bumpers and sill (rocker) panels were fitted, plus twin headlights, and under the bonnet (hood) was the Spica mechanical fuel injection system.

Right - The dashboard of a 1982 Alfetta saloon, showing the angled glove compartment and instrument binnacle, which looks rather 'plonked' in place. It's fairly well co-ordinated, though, and there's a matching wood-rim wheel and 'golf club' gearlever knob.

two-tone cloth and an instrument binnacle with semi-circular speedo and rev-counter, plus auxiliary dials and warning lights, ahead of the leather-rim wheel. By the time the 116-series Giulietta went out of production in 1985, the instrument panel had metamorphosed into a broad oval binnacle, with the principal gauges to the left and centre, and a battery of warning light symbols to the right. In 1979 the 1.8- and 2.0-litre versions appeared, and, like the Alfetta, used 1779cc and 1962cc engines.

Successive revisions of the 116-series type number saw the 2.0-litre Giulietta appear as the Super from 1980, the Nuova L from 1981, the Ti in 1982, and simply the Giulietta 2.0 from 1984. The 2.0 Ti was a limited edition model which differed in having

Bereft of its front bumper and in lowered race trim, the angular Giulietta shape begins to work aesthetically. Here is midfield runner Andy Page in his turbocharged 2.0-litre Class A car at Lydden Hill in May 1999.

Like military camouflage, Steven Griffin's geometric livery makes the shape of his Class E Giulietta harder to read. Seen here at Snetterton in 1999, the 2.0 Giulietta runs on retro-fit Revolution alloy wheels.

Rear view of the Alfetta 2000 in its final Gold Cloverleaf guise, showing the gurney spoiler at the trailing edge of the bootlid. It also has the finned Campagnolo alloy wheels, while some cars were supplied with the plainer dished variety.

grey lower quarters and matt black rubbing strip. There was no brightwork surrounding the front and rear screens, and the window glass was tinted bronze. Inside, the seats were upholstered in leather and cloth fabrics, with a grey leather-rimmed steering wheel.

Range topper

The top-of-the-range Super was available only in metallic grey, and was distinguished from the smaller-engined models by a beige stripe that ran the length of the waistline, and the word 'Super' emblazoned in stencilled graphics below the A-pillar. Like the Ti, the Super ran on contemporary dished Campagnolo alloys, an improvement on the base model's although the finned Campagnolos fitted in 1984 were aesthetically more interesting. All Giuliettas ran on 13in wheels, whereas the Alfetta used 14in wheels.

The first styling revision for the Giulietta came in 1981, with the lower third of the bodyshell painted a different colour and rubbing strips all round its body at bumper level. The front valance became accentuated into an air dam shape, incorporating driving lights and a headlight wash-wipe system. The facelift of 1984 included new bumpers with a more complex profile, together with a different shaped front valance and side protectors. The instrument binnacle was now in its semi-circular housing, while the centre console was similarly rounded off to match. The switchgear was relocated to the centre of the dash, and more thought was given to storage bins.

For just two years, between 1983 and 1985, the Giulietta was equipped with the 2.0 VM Turbodiesel engine, and was identifiable in having extra air intakes. Apart from extra sound dead-

ening material, the diesel was the same as the petrol-engined cars.

The 2.0 Turbodelta of 1983 was a different matter. Developed by Autodelta as a potential successor to the GTV6 in European Touring Car Racing, where its saloon car status offered more promotional potential than the coupé, the idea was to produce 1000 units for homologation purposes. The Giulietta Turbodelta was, if anything, more refined than the Alfetta version. Whilst under development, it used an Alfa Avio turbo, allied to two twin-choke carbs, but when the car went on sale it came with a KKK unit. Exhaust valves were sodium-filled with stellite seats, and other additions included an oil cooler and ventilated disc brakes. Along with uprated suspension, the Turbodelta used five-stud hubs with 'telephone dial' alloy wheels, and was fitted with a boost gauge and nice Momo steering wheel. All were metallic black with a red stripe along the car's midriff. However, we never heard much about it because the project was stymied by the management's reluctance to give it full backing, preparing instead to put its weight behind the Alfa 75 scheduled for launch in 1985. In the end only 361 units of the Turbodelta were produced - and sold off cheaply.

With regard to the rest of the Giulietta series, production of the 1.3-

Alongside the Alfetta saloon was manufactured the Alfetta GT, which by 1980 had evolved through 1.6-, 1.8-, 2.0-, and 2.5-litre versions. This is the latter V6-engined GTV6 model, which I raced in the AROC series, and captured on film by Mary Harvey at Castle Combe in 1990. In the hands of Andy Rouse, the GTV6 won the European Touring Car title in 1983.

Below, left - The Giulietta was designed by Rudolf Hruska along the still-current wedge-shape lines and utilising the powertrain and suspension from the Alfetta. It was introduced in 1977.

This atmospheric shot of the Giulietta shows the 122bhp 1.8 version around the time of its launch in April 1979. There are no rubbing strips along the sides, C-pillar vents are body-coloured - note the filler cap - and, whilst it has finned Campagnolo alloys, there are no headlight wipers yet.

engined car stopped in 1984, and the rest of the range went out of production in November 1985. In total, 560,636 Alfetta saloons were produced, and 348,161 Giuliettas of all versions.

The Barge

If this sub-heading seems downright derisory, it's not meant that way; but most people seem to refer to the Alfa 6 by this epithet. It has, justifiably, a hard core of devotees, who perceive in its ample proportions and lusty V6 powerplant the contents of a luxurious touring car, bordering on a limo. From the side, at least, it was a handsome beast. Alfa Romeo's head of design at the time was Carlo Bianchi Anderloni, who had worked at Arese since 1966 when Touring Superleggera folded, and he was on record as being quite content with the Alfa 6's styling. Besides which, I'm a great fan of barges, with their quite majestic broad planes of metal and understated power.

The Alfa 6 is arguably no different in concept to the big 6C 2500 Berlinas produced by Touring and Farina in 1949, and, in fact, Alfa has come out with large saloons at various times in its postwar history. They were what we now describe as 'executive' class saloons, and have with varying degrees of success fulfilled the role of flagship in the Alfa range. In the late 1950s and early 1960s, the 2000 and 2600 Berlinas served that purpose, with

Top-of-the-range Giulietta was the Super, which came in metallic grey only and was identified by the logo at the base of the A-post and a beige stripe that ran the length of the car's waistline.

Below, right - With the 1981 facelift, the 1.3- and 1.6-litre models received the rubbing strips along the sides, in contrast to the 1.8 model that got broader plastic side mouldings.

moderate accomplishment.

Designated the Tipo 119 A, the Alfa 6 came out in 1979, although it was originally intended to launch ahead of the Alfetta, which is some delay! Had it appeared on schedule some four or five years sooner, there's every reason to suppose it would have done a lot better. For a start, the design would have been fresher, and its 2.5-litre six-cylinder powertrain would have seemed even more *avant garde* than it did in 1979. The 60-degree V6 engine breathed through six single-choke downdraught Dell'Orto FRPA 40 carbs, with 12 valves actuated by a single camshaft in each bank, and developed 160bhp, making it the fastest Italian saloon on the market and the most powerful 2.5-litre four-door saloon in the world.

Since the Alfetta came out first, we'll say that the Alfa 6 was based on the Alfetta platform, complete with deformable structures, and longer wheelbase and greater overhangs front and rear. Like the Alfetta and Giulietta models, it used the transaxle gearbox arrangement. Suspension was also based on the Alfetta's De Dion rear axle with Watts linkage, floating halfshafts and inboard disc brakes and anti-roll bar. Up front were the now-familiar torsion bars, castor rods, wishbones coil-springs and dampers and anti-roll bar. Cars destined for the UK were supplied with ZF three-speed automatic transmission that used a hydromatic torque converter.

In 1983 the model was facelifted in a move that eliminated the rubber

It's not unusual for Alfa owners to swap engines, either because it's possibly a cheaper way to go than having a major rebuild if the old unit's clapped out, or simply to gain extra performance by fitting a bigger unit. But in this case, the Milanese owner has fitted a V6 unit in his Giulietta and it looks like a very neat installation.

The Alfa 6 was finally unveiled in 1979, some five years later than had originally been intended. It was based on the Alfetta floorpan, originally equipped with twin headlights, and was the first Alfa Romeo to receive the aluminium V6 engine.

Below, left - Although the Alfa 6 has been criticised for looking like nothing more than an overgrown Alfetta, it was actually a handsome car, in this, its original format, uncorrupted by plastic add-ons. The C-pillar vents are in black, incorporating the fuel filler on the right, and bumpers are predominantly chrome.

This cutaway illustration of the Alfa 6 in its original format, by Bruno Betti, reveals the 2.5-litre V6 engine and gearbox unit, Alfetta type suspension front and rear, and the car's voluminous boot.

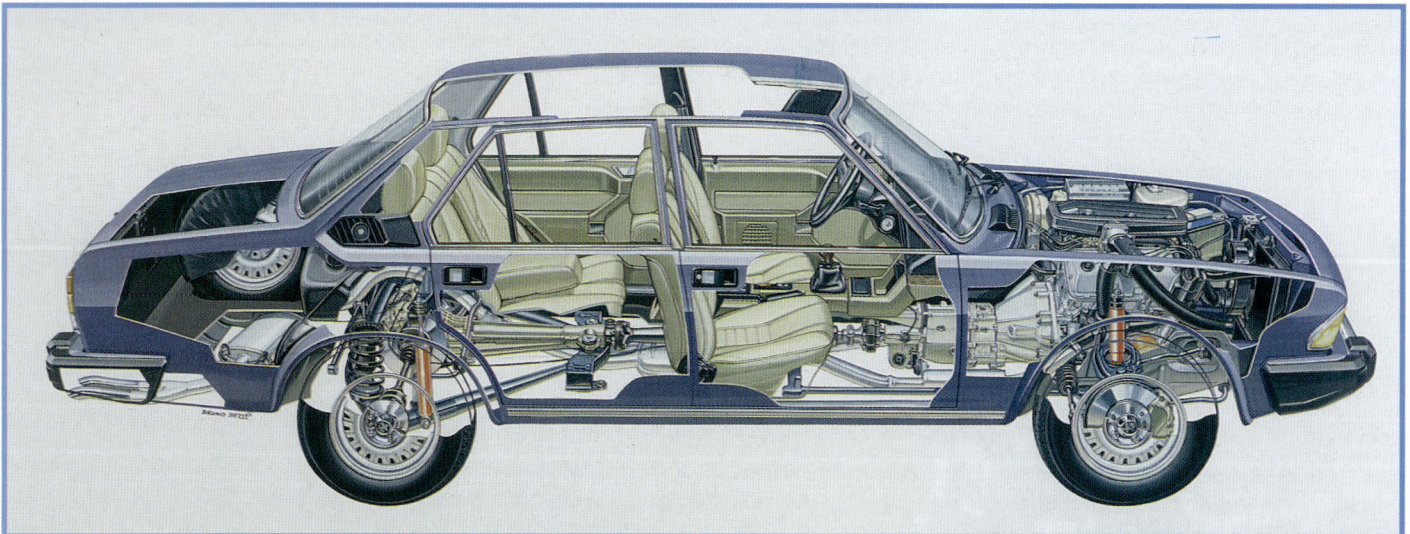

overriders on the bumpers and installed rectangular headlights, which served to bolster the car's already angular image. It was given a fresh rubbing strip along the flanks at bumper level this time, revised air vents on the C-pillars, and dished alloys replaced the finned sort on the earlier car. Let's not beat about the bush about the original dashboard; it was hideous, a total mish-mash of materials and unrelated architectural shapes, topped by a steering wheel that would have looked more at home on the bridge of the Starship Enterprise. With the facelift, new seats and upholstery were fitted, along with a redesigned dashboard - which was something of a mercy. On the mechanical side, the arsenal of carburettors was replaced with a Bosch L-Jetronic fuel injection system annexed from the GTV6. This had the instant effect of improving fuel consumption and emissions quite dra-

Operators on the Arese assembly line carefully lower a V6 unit into the engine bay of an Alfa 6 using a pneumatic hoist. The car's front wings are protected against scuffs. Its build ticket appears to be lodged in the radiator grille, and the yellow headlights suggest it's bound for France.

Middle right - The facelifted 1983 model was given rectangular head-lights, side rubbing strips and a front air-dam with splitter. Road wheels changed to a flatter dished-pattern alloy type.

Bottom right - The revamped Alfa 6 was given heavier rubber-clad bumpers that were fared in towards the rear wheelarches, and the plastic moulding on the C-pillar was re-vised. The total effect was to muddle what had been a clean, if unremark-able, design.

matically at the sacrifice of very little in the way of performance. Rather con-trary results were obtained when the 2.5-litre unit was downsized to 2.0-litres for a tax-cutting version intro-duced in 1983. Not only was it under-powered, it was also less economical than the 2.5-litre version, and air con-ditioning was optional rather than standard issue.

There was another version of the Alfa 6 and that was the 2.5-litre Turbodiesel, introduced in 1983 in the guise of the facelifted model. For this car, a brand-new five-cylinder, in-line turbocharged unit was specially built by VM, developing 105bhp at 4300rpm and enabling it to reach a top speed of 170kph. Alfa 6 production totalled 12,260, with a further 2630 turbodiesels made, and all three ver-sions went out of production in 1985.

The right angle

When Alfa Romeo attempted to break into the luxury executive market in 1984, the weapon it chose was the Alfa 90, basically the Alfetta saloon which had been given the Bertone treatment to bring the ten-year old design up to date. Included in the 90's specification

The dashboard of the Alfa 6 was something of a disaster, with a mish-mash of geometric forms vying for attention with miss-matched materials. Even the steering wheel spokes have fake wood inserts. And yet, if you disregarded all that - like, wore a blindfold - it was an excellent, relaxed, long-distance touring machine.

was a self-adjusting front spoiler, controlled by a small hydraulic damper, which was of dubious use except perhaps on the *autostrada*. The car deserved to do better than it actually did, but the unfortunate 90 found itself overshadowed by the more complex aesthetics of the incoming 75, launched in 1986.

As an evolved Alfetta, the 90 shared most of the suspension and running gear of the outgoing saloon, including De Dion rear suspension, inboard discs and transaxle arrangement with coil springs and dampers at the back, and the same wishbone, torsion bar and castor rod set-up at the front. Ventilated discs were fitted at the front,

along with rack and pinion steering. The base model 90 was powered by the 1779cc engine, and the middle-of-the-road sector was catered for by the 2.0-litre model, which could be specified with carbs or fuel injection as the 2.0i. In a bid to increase flagging demand, the 1.8 was given a performance hike in 1986 that improved top gear acceleration, gaining Super status accordingly, while the 2.0-litre model also became the Super, fitted with injection only. A plainer front grille, without the benefit of horizontal lines, was fitted, and the self-adjusting spoiler was done away with. The 90's appearance was always going to be compromised by its massive plastic bumpers, but the earlier grille at least broke up the expanse of black. The rear numberplate was inset into the bootlid, which was an unusual styling quirk.

The interior was already quite some way nicer than that of the Alfa 6, with neatly stacked centre console and instrument binnacle and, as a carrot to tempt the executive owner, a cost-optional removable briefcase fitted under the glove compartment. From 1986 new velour-covered seats were installed and, curiously, the speedo and rev-counter worked on an inclined linear plain, going upwards from left to right. The heater controls were the

The Alfetta and its derivatives proved they handled and performed well, and here's an Alfa 90 being put through its paces by Tony Phillipson at an AROC Autotest at North Weald airfield in Essex in 1987. This 90 started life as an Alfa GB press car.

The Alfa 90 could be considered a stopgap as it bridged the gap between Alfetta and 75, yet it does stand as a model in its own right, since, like the 1750 and 2000 Berlinas, it was the product of Bertone's design studio. The brief was to update the Alfetta, based on the existing floorpan and use of standard components, and that's what you got.

same as those fitted in the new Alfa 75 that came out in 1985.

The range-topping model was the 156bhp Alfa 90 2.5 *Quadrifoglio Oro iniezione*, powered by the V6 engine from the Alfa 6 and GTV6. This unit was of all-aluminium construction, with 9.0:1 compression ratio and bore and stroke of 88mm x 68.3mm, giving 2492cc. Maximum power was 156bhp at 5600rpm, and torque 154.9lb/ft at 4000rpm. Some will argue that its most distinctive feature is the glorious bark it emits, wailing in a crescendo as it hits high revs, and which, in the case of the Alfa 90, makes the engine more seductive than the car it's powering.

This car was in production from 1984 to 1986, superseded by the 90 Super 2.5, in line with the facelifted four-cylinder models in 1986. Technical specification differed only in so far as the V6 cars had ventilated front discs, with optional ABS, and a different final drive ratio.

Another two variations on the theme appeared in 1986. The first was the 1996cc V6 engined car in 2.0-litre format, a familiar ruse that pitched the model under the 2000cc VAT threshold, saving the 38% levied on such gas guzzlers in Italy. This model developed only 132bhp at 5600rpm, but at least it made the right noises. Unsurprisingly, the Alfa 90 was also

fitted with the 110bhp 2.4-litre VM turbodiesel, which was capable of 180kph. The 90 range was phased out in November 1987 in deference to the 164, at which time a total of 56,428 units had been produced.

Your number's up

A new model was unveiled in 1985 to mark the company's 75th anniversary, and appropriately it was called the 75. But why didn't Alfa give it a name, especially since it was essentially a descendant of the Alfetta and Giulietta? In 1998 Rover launched its 75 model, and I was tempted to suggest it was numbered after the Alfa. But, of course, there was a venerable precedent in that case.

Like the 90, the 75 was designated a tipo 162 series car, and it had the same 2510cm wheelbase as the 90 and Giulietta, with less front and rear overhang than the 90, and slightly more than the Giulietta. Stylistically, the 75 developed some of the themes begun in the Giulietta, but carried them off better by caricaturing them: the wedge-profile was accentuated by a plastic trim piece that ran the length of the car's waist and kicked upwards from the C-pillar aft; the grille was made more interesting simply because it slanted outwards at the sides instead of being rectangular like on the

Giulietta, and the front corners were better rounded off. The 75's compact proportions disguised a capacious boot, which extended underneath the rear window to provide 17.5 cubic feet of space. The shape worked better in a dark colour scheme as the dark colours camouflaged the heavy bumpers and plastic side strip.

When introduced the 75 was available with almost the entire range of twin-cam engines, including the 1570cc, 1779cc and 1962cc units, but no longer the 1357cc that had been available in the Giulietta. As with its immediate predecessors, the Alfa 75 was a rear-wheel drive car, with transaxle gearbox and De Dion rear suspension, inboard rear brakes and the same front suspension as well. The base model Alfa 75 was initially equipped with carbs, which, by 1989, had been replaced by Bosch Motronic ML 4.1 fuel injection, with a catalytic converter fitted the following year. Also in 1990, the Alfa name was dropped from the car's title - it was simply the 75 1.6 IE.

Inside, the design quirks for which Alfa has always been renowned were present in the U-shaped handbrake lever that hinged up over the centre armrest/dump bin, and the electric window controls, which were in an unlikely location above the windscreen by the rear view mirror bracket, along with the interior light and map reading light. You just had to live with the car to get used to that one. Another strange innovation that I first came across in my mother-in-law's 75 1.8 was a vertical strip of yellow lights set in the right-

Not only was the 90 available with the same engine range as the forth-coming 75 (bar the 1.6), its interior prepared you for some of the weird ergonomics that were in the pipeline. On the 90 Alfa tried out the same type of robust plastic fascia, steering wheel and door furniture, similar seat upholstery, and the 90's instrument panel featured a diagonal pairing of linear speedo and rev-counter.

The 90 was available with 1.8- and 2.0-litre twin-cam engines, in the latter case with either carburettors or fuel injection, as well as the 2.5-litre V6. It also had a self-adjusting front spoiler that was controlled by a miniature hydraulic damper and became more prominent the faster you went.

hand side of the instrument panel that lit up in sequence to advise you when you should be changing gear. The 75's driving position was a little better than its forebears, and was enhanced by a steering wheel now adjustable for reach as well as rake, although the splayed knees attitude still prevailed.

The 75's plain front grille was col-our coded in a 1986 facelift, at which point Alfa brought out a turbocharged version of the 1.8 known as the Alfa 75 1.8i Turbo (the US market got the higher trim spec Turbo America desig-nation, but it was never imported into the UK), with forced induction based on an intercooled Garrett T3 system. Externally the 165bhp turbo model differed in having slightly wider wheelarch flares, wider wheels and body-coloured bumpers. From 1987, the model suffered the indignity of those massive front and rear bumpers, designed to comply with US safety

Launched in 1985, the Alfa 75 was named (or numbered) in celebration of the company's three-quarters of a century in business. The Alfa shield had by now become compressed to such an extent that it bore little relation to the refined examples of the postwar years.

Minette Rice-Edwards in the Alfatech 75 3.0-litre V6 at the tight little Lydden Hill circuit, Kent, during a round of the AROC Championship in 1999. She also competed with an Alfetta GTV in 1998.

Engine bay of a 1.8-engined Alfa 75. In normally aspirated format the engine produced 120bhp at 5300rpm, 165bhp at 5800rpm with turbo, and 280bhp at 5800rpm in turbocharged Evoluzione racing format.

regulations but which added aesthetically undesirable bulk to both ends of the car. US models were sold under the nomenclature of the 75 Milano, in either Gold, Silver or Platinum Cloverleaf form with varying degrees of trim options and equipment levels. The *Quadrifoglio Platino* , for instance, offered ABS and air conditioning.

Extreme evolution

The 75 appeared in its most extreme configuration as the Turbo *Evoluzione* the same year, a 280bhp device homologated for competition use, and from which the Veloce body kit subsequently seen on other renditions of the 75 was derived. Racing versions used the KKK turbo, allied to an intercooler and fuel injection, and the model was developed into a competitive Touring Car under the direction of Alfa Corse's competitions manager Giorgio Pianta, who had been a member of the Autodelta GTA squad in the 1960s and Lancia rally chief in the 1970s. The cars were hugely successful in the Italian and European Touring Car

The 75's cockpit may pose some curious ergonomic mysteries, like the spade-handle handbrake lever between the seats, or the electric window switches up by the rear view mirror. But if you live with it long enough, these seem perfectly acceptable traits. Most irksome about the 75 controls was the warning lights that came on when there wasn't a problem - like, suggesting an oil crisis on the motorway when in reality the sump (oil pan) was full.

competitions on race circuits and events like the Tour of Italy, won by a 75 in 1989. The Alfa Corse *ad hoc* squad included such luminaries and legends as Jacques Lafitte, Riccardo Patrese, Sandro Nanini, Jean-Louis Schlesser, Michael Andretti, Nicola Larini, Thomas Lindstrom, Gabriele Tarquini, Rinaldo Drovandi, Paulo Barilla, Gianni Morbidelli, Gianni Giudici, and Giorgio Francia. In 1991, Luis Perez Sala was Spanish Touring Car champ in a 75. In the UK, the earliest exponent of 75 racing was the evergreen Jon Dooley, who, along with Rob Kirby, ran the Dealer Team cars

with John West Salmon sponsorship. Even at club racing level the 75 was prominent. When I dipped my toe into the choppy waters of the AROC Challenge in 1990, the series winner was Roger Kay in his 75 3.0 V6, racking up 12 outright wins and setting 10 lap records in the process. The car was prepared by York-based Alfa dealer Alwyn Kershaw who had performed a similar job on Kay's GTV6 the year before, and much of the car's excellence stemmed from its blueprinted V6 engine.

By the end of the decade the 75 was still being campaigned in the UK

AROC series, and one of the most successful runners was Graham Presley, whose 75 was powered by the 1.8-litre turbo engine prepared by BLS Automotive of Lincoln.

One spark good, two sparks better

Unlike the contemporary Alfa 33, there was never an official Giardinetta version of the 75, although a prototype was created by Rayton Fissore for the Geneva Show in March 1987. It looked like an overgrown 33 Sport Wagon. However, this particular Geneva Show was the setting for an altogether more momentous announcement.

At launch, the 2.0-litre 75 had been fitted with the regular 128bhp unit that had seen service in the Alfetta and Giulietta. Now, though, Alfa played a significant trump card when it unveiled the 148bhp 2.0 Twin Spark version, which instantly moved the 75 onto a higher plane than previous models in this segment. It was now reliable, economical and faster, and the power delivery much smoother and consistent. The traditional guttural rasp of the old twin-cam was sublimated to more of a purring growl. The Twin Spark engine ran with the Bosch ME7 Motronic electronic management system that controlled the ignition and fuel injection, and the two-plugs-per-cylinder made famous in the GTA racing engines. Alfa Romeo had employed two spark plugs per cylinder as far back as 1923, of course, and in the context of the 75, the advantage was more efficient combustion without resorting to the complexity of multi-valve

The Carabinieri Giulia Supers which featured in that wonderful Michael Caine movie The Italian Job *were typical of their day. In real life they probably wouldn't have appeared so incompetent, and it's perhaps a good thing that the British police force never picked up on the 75, pictured here during evaluation trials.*

This Trofeo Garage-sponsored 3.5-litre 75, driven by AROC Championship front-runner Alan Marshall, sports the full Veloce body kit plus an outsize rear wing. Alan's big-engined car was running at Spa Francorchamps in May 1999.

cylinder heads. The Twin Spark's second distributor was neatly mounted on the front of the engine ahead of the oil filler cap. In catalysed form, the 2.0 Twin Spark was known as the Europa, and performance lagged slightly because of the overly complex exhaust system.

The Twin Spark interior was an altogether jollier environment than anything seen previously, featuring speckled herringbone tweed-like upholstery with a hint of red and orange, either in two-tone cloth or with vinyl inserts. The front seats were supportive Recaro style, with adjustment for height and rake of the seat squab as well as rake of the backrest. Having said that, the cubic dashboard architecture was stacked up as haphazardly as the speaker system at a village hall gig.

Exterior modifications in 1988 included a small spoiler on the front valence, matched by a similar protuberance on the rear valence and a little

Waiting their turn in the Brands Hatch paddock at the British GP meeting in July 1986 are the Alfa Romeo Dealer Team Alfa 75 2.5 V6 racers of Rob Kirby (30) and Jon Dooley (31). In the background, AROC Secretary and team manager Michael Lindsay discusses tactics with driver Jon Dooley. Rob came 2nd in class while Jon retired.

The works Alfa 75s were campaigned in the late-80s with the 280bhp turbocharged 1.8 engine, and this is the formula that Graham Presley used to great effect several years running in the AROC series with his BLS-prepared Class A 75. The car waits alongside its stablemate 33 in the damp Donington paddock in March 1999.

gurney spoiler on the trailing edge of the bootlid, made of the same rubberised plastic as the side strips. The sills were treated to matching mouldings that hinted at racing side skirts, and the race-rep Veloce body kit took these cosmetic trim details a stage further.

Suspension tweak

In 1989 the Tiplers bought one of these cars from Alfa GB's press fleet and it wasn't long before the dark metallic blue 75 Twin Spark had received a handling kit from Derbyshire-based suspension guru and erstwhile Autodelta driver Rhoddy Harvey-Bailey. As I interviewed him about his Autodelta experiences for my book on the Giulia Sprint GT and GTA, his colleague fitted the handling kit. This consisted of a thicker anti-roll bar at the front, stiffer springs and new Koni dampers. Curiously enough, they didn't advocate lowering it, and although it sharpened up the ride, handling and turn-in, my subsequent experience with other models leads me to believe

that dropping it by an inch and a half would have been a good idea. Circumstances rather brought any further experimentation with this particular car to a premature end. Returning home late one night after a nocturnal photo shoot involving an Audi quattro, I flipped the 75 on the back road between Chepstow and Tintern, and nearly ended up in the River Wye. It got away from me on a corner - diesel spillage or something, Officer - and collided with a cliff, which was enough to turn it over. I was shaken but unscathed, although my normally placid old dog was a bit put out. Not a panel of the 75 was left undamaged, and an older car would have been written off. But it was repaired to as-new condition and was sold soon afterwards when we went off to Holland to live on a barge.

The fastest - though not necessarily the most agile - model in the 75 line-up was the V6-engined model. This started life as the 156bhp 2.5-litre V6 Green Cloverleaf in 1986, which used

the now-familiar six-cylinder engine from the GTV6 and 90, complete with Bosch Jetronic fuel injection. Work had started on this engine as far back as 1971, and it had been intended as a replacement for the regular in-line twin-cam four. Development was carried out under the direction of Edo Masoni in Alfa's engine development department. Masoni's first idea was to adapt the V8 unit from the original T33 sports prototype and Montreal coupé, but this was clearly going to be too complex for a family car. The V6 was therefore a brand new design, with cylinder banks set at 60 degrees and incorporating narrow-angle valves and a single hydraulically tensioned belt-driven camshaft to each bank. The 2.5 V6 engined model was also available with automatic transmission. In 1987, the V6 capacity was upped to 3.0-litres and a limited slip differential was fitted. Examples shipped to the States were fitted with catalytic converters and the model was known as the 75 Milano 3.0i *Verde*. Apart from better pulling

When Alfa Romeo introduced the Twin Spark engine at the Geneva Show in 1987, the venerable twin-cam gained a new and lusty lease of life. Incorporating a second spark-plug for each cylinder improved combustion significantly, so that driveability was improved through better throttle response, and power was up to 148bhp. This one positively purred, whereas the old one growled.

power and higher 137mph top speed of the 3.0-litre car, it differed in having a smaller boot space than did the Twin Spark because its larger fuel tank took up space behind the rear seats.

Family motoring

My own 75 V6 was a black 3.0-litre Veloce model which I swapped for my newly-restored GTV6, based on the need for a four-door family saloon. The 75 was a bit quicker than the GTV6, but the coupé still had its race suspension set-up, along with BF Goodrich Comp TA tyres and competition brake pads, so it was way superior to the 75 in those respects. It wasn't long before Norwich-based Alfa specialist Richard Drake had lowered the suspension and fitted a new set of Spax dampers, and around the same time I discovered Yokohama tyres. These 195/55 x 15 low profiles were not only keenly priced at about £60 a cover, but made a significant difference to the 75's roadholding. Now it really could be thrown around in a hooligan sort of a way. And despite what they say about the Twin Spark being lighter, and therefore less nose-heavy than the V6, I reckon there's no substitute for a well set up six-pot screamer. It delivers enough power coming out of a turn, and just keeps on going, while the 2.0-

litre four is running out of breath. The one serious deficit of the 75 V6 3.0i was its appalling lack of brakes; half-a-dozen laps of Snetterton on a club practice evening and the brakes were virtually shot. Still, I suppose that saves the engine from being roasted when this happens as there's no realistic option but to stop and let the brakes recover.

When called upon to perform the function for which it was designed, that's to say transporting the family and its luggage, the 75 acquitted itself admirably. I have family in Cumbria, so trips from Norfolk up the A1 and through the Dales were regular events - a good mix of fast A-roads (when driven at night) and startling B-roads. Once it transported six of us - that's two teenagers, two littles and two adults, plus baggage - all the way from Spain and up through France after a Eurocamp holiday. This may be seen as irresponsible by safety adherents, but I record the journey simply to demonstrate that it could be done - and no-one would have thought twice about it in the days before seatbelts. There were times when the car's lowered stance annoyed, such as the increased vulnerability of the splitter on the front air dam, which came into contact with kerbs now and then. It

was such scrapes that showed the body kit for the lash-up it was. At the front at least there was no discernible bumper behind the plastic, and the whole ensemble, air dam and 'bumper', was attached to the car in the most rudimentary way. Several different methods of fibreglassing, drilling and wiring up were tried in a bid to keep it in place, which were ultimately only partially effective. Also, the car's lowered ride height caused periodic altercations between the V6's twin down-pipes and 'sleeping policemen' (speed humps), and grounding was a risk when driving fast along undulating rollercoaster roads. I wouldn't have had it any other way, though.

Do you really want to know about the Alfa 75 2.0 Turbodiesel? I thought as much. But just for the record, the model was in production from 1985 to 1991, when the model range was phased out. Equipped with the necessary sound-deadening material, the TD used the 95bhp four-cylinder VM unit until 1986 when it inherited the 110bhp 2.4-litre version from the Alfa 90 Turbodiesel. It was never a temptation that British Alfa owners had to agonise over, whereas in Continental Europe there were fiscal advantages to running such a car. They made 47,962 turbodiesel-powered 75s, out of a total build of 364,661 of all types of 75. At 9526 units, the 3.0-litre V6 model is the rarest.

Monster fun

On the other hand, the 75 did have the distinction of providing a floorpan for the outrageous ES 30, better known as

Here's one that didn't owe me any favours. My 1988 3.0-litre 75 V6 had done over 150,000 miles by the time it was ten years old, and was, mechanically, at least, supremely reliable. Its negative camber settings used to knock out the front Yokohamas a bit prematurely, and the front apron used to fall off from time to time, but in general it was an excellent car.

From 1987, the quickest version of the 75 was powered by the 2959cc version of the 60-degree all-aluminium V6 engine, producing 192bhp at 6000rpm. It was said to emit the most wondrous noise this side of a Ferrari, and who's arguing?

Customising can get out of hand, but this 75 Twin Spark belonging to John Dack is a good example of how far you can reasonably go without looking too leery or ridiculous. It features a four-headlight grille with special front apron, giant 19in diameter Supertouring wheels, and a leather interior as exotic as the façade.

Each generation of Alfa has its wild child, and the coupé derived from the 75 was the Zagato-made ES 30, better known as the SZ. It was originally intended as a pivotal model which would herald the Alfa Romeo renaissance, but the resignation of its chief proponent, Vittorio Ghidella, to some extent reduced the impetus to produce it.

the Zagato designed and built SZ coupé made between 1989 and 1991. This six-headlight stunner was dubbed *il Mostro*, and had a mixture of Modar and glassfibre exterior panels bonded to its steel chassis, with powertrain sourced from the 3.0-litre 75, albeit tweaked to 210bhp. The rose-jointed suspension was derived from Alfa Corse's racing experience, and included a 40mm hydraulic ride-height adjustment on the shock absorbers so you could have it with almost no ground clearance, or just a bit. Always a two-seater, the SZ was hailed by some motoring scribes as the best car they'd ever driven. Such adulation, plus a limited production run - 1035 units -

111

From the start the 75 was offered with a turbodiesel engine. The VM-produced 2.0-litre unit was fitted until 1986, when superseded by the 2.4-litre version. The unit featured a gear-driven valve-train, parallel overhead valves and an Alfa Romeo Avio turbocharger with air-to-air intercooler and wastegate.

brought speculators into the market and values rocketed. A convertible version - the R.Z. or Roadster Zagato - was perhaps even more desirable, as is the way with sports cars, and 241 of these were made between 1992 and 1993. Now, at the turn of the century, you can get an SZ for around £20k. But you can also get a body-kitted 75 for a tenth of that, and have four seats into the bargain, so I wonder which is better value?

Blown over

If you want still more performance from your 75, there's a Biggleswade-based firm called Schwaben Garage which offers a range of conversions based on the V6 engine, ranging from Twin Turbo to Supercharging. The Twin Turbo set-up produces a stonking 280bhp, which will set you back some £3650, while the supercharged route costs about £4250. Lesser conversions include 24-valve engines and fitting high-lift cams. If you're wedded to your 75, that's a way to go. Otherwise it's a lot of money to spend compared with the resale value of the cars these days.

Apart from four-wheel drive models like the 155Q4 and 164 Q4, the 75 was the last rear-wheel drive car that Alfa made, and this was something mourned by purists and driving enthusiasts, since front-drive cars lack the fluency and traction of rear-wheel drive. And therein lies its epitaph, really. It was an odd-looking car, much underrated, and may not have been as quick point-to-point as a front-drive 155 or whatever. But it was a spirited, practical car and it was ace to drive.

Unveiled at the 1989 Geneva Show - on the Zagato stand - the SZ was hailed by some as the most exciting car they'd ever driven. The cockpit certainly invited such a prognosis and, with only 1035 units built, the SZ was destined for instant classic status.

9

SOUTHERN OUTPUT

The previous chapter dealt with the Alfa Nord rear-drive cars beginning with and descended from the Alfetta, which was virtually contemporary with the 'Sud. Now we've got to go back a decade or so to pick up on what was going on in Alfa Romeo's southern plant in the wake of the Alfasud.

In 1979, the company's finances weren't looking too bright, and a joint venture with another company was mooted, with a view to symbiotic benefits. Thus it was that, in 1980, Alfa Romeo went into partnership with Nissan: the Japanese manufacturer being keen to establish a believable foothold in Europe. The advantage to Alfa was that Nissan would fund the production of a new model based on the tried and tested componentry of the Alfasud, installed in the equally

well time-served Nissan Cherry bodyshell, which was shipped over in panels from Japan. The car's grille bore the Alfa shield and logo, and it was called the Arna, an acronym of Alfa Romeo Nissan Automobili, and designated the *Tipo* 920A. It was assembled on a brand-new largely robotised production line at Pianodardine, Avellino, and unveiled in November 1983.

Arna good

Now I'm not prejudiced against Japanese vehicles - I've had a Honda CBR600 motorcycle in the past - but I can't imagine why anyone vaguely interested in Alfa Romeos would want to buy one with a bland Japanese body. Maybe if you were a fan of Nissans, but wanted something a little

The Arna was conceived at a time when Alfa Romeo was in financial peril, and one way of consolidating its affairs was to embark on a joint venture with Nissan. The result of this coalition was a Cherry with Alfasud mechanicals.

Because of its mixed parentage, the Arna is not greatly loved in Alfa Romeo circles. However, Dave Streather demonstrated that it could be every bit as good as an Alfasud or a 33 by winning the AROC Championship in 1997 and 1998. The car continued to be campaigned to good effect by Shane Crumpler in 1999.

different you'd go for an Arna? We know that cars share platforms - like the Croma/9000/Thema/164 Type 4 models - but it seems to me that, in this case, it would be like buying, say, a Honda 650 Bros that was fitted with a Ducati 600's engine and front suspension. It appears that most *Alfisti* agreed, and shied away from the Arna in droves. The only people to have done anything useful with one - in the UK, at least - are Dave Streather, who won the UK's AROC Championship in 1997 and '98 by virtue of accumulated Class C wins, and Shane Crumpler who bought the car. Actually, there's no particular secret to setting up an Arna. Once the Nissan rear suspension set-up with its drum brakes is sorted out, it's essentially the same as an Alfasud, albeit with the Japanese shell.

The Arna was available in three- or five-door versions, and the base model was powered by the 1186cc flat four unit, which gained breakerless electronic ignition in 1984. That year, an SL version was introduced with metallic paint and optional sunroof. One thing that could be said in its favour was that access to the boot was good, thanks to the huge rear hatch that doubled as the rear screen. All models had a rear wash-wipe system.

Just weeks after the original Arna was launched, the Arna ti appeared, and you can't help but view it as a placebo for the indignant Alfisti who wondered what on earth the company was playing at. This model was fitted with the 86bhp 1351cc boxer unit from the 'Sud ti and Sprint series, and its three-door shell was distinguished by a tiny chin spoiler up front and a vestigial wing on the trailing edge of the rear hatch. Matt black bumpers were accompanied by black wheel centres, and inside the Arna ti had a rev-counter and jazzy seat upholstery.

Strangely, the three-door 1.5-litre twin-carb version was made in right-hand drive only, and marketed as a Nissan Cherry Europe. The joint venture with Nissan drew to a hasty close in 1987 as Alfa Romeo emerged from the titanic corporate struggle between Ford and Fiat to secure its acquisition, the future now apparently assured under the comforting umbrella of Fiat ownership. A total of 46,231 Arnas were made in three years.

Bestseller

We've jumped ahead of ourselves a little bit and need to backtrack to the uncertain times of the early '80s to check out the genesis of Alfa's most prolific model to date, the Alfa 33.

The Alfasud was a hard act to follow, and Rudolf Hruska's design for the 33 didn't succeed in matching the cheeky image projected by the 'Sud. Alfa tried to give the new model a fighting chance by numbering it after the exciting Tipo 33 sports prototype with which the company won the World

The 1351cc Alfa 33 was launched in Spring 1983, and its relatively subdued lines owed more to the larger Giulietta than its actual predecessor, the Alfasud. Under the skin was much the same powertrain and suspension layout as had the front-drive 'Sud.

Championship for Makes in 1975 and 1977. But it's a moot point whether many customers registered the connection. In any case, the 33 was designated the Tipo 905 in Alfa's complex and constantly changing type numbering system. The 33 went on to be Alfa Romeo's best selling model, with 1,217,026 units built during its 11-year history. Surprisingly, this was 310,000 more than the Alfasud, which was in production for a year longer.

Whilst retaining the same drivetrain and wheelbase as its predecessor, the 33 was a rather larger vehicle by virtue of greater overhangs front and rear, although it was a little bit lower in overall height. Stylistically, the 33 owed rather more to the rear-drive Giulietta than the front-drive 'Sud, and its C-pillar in particular an-

By June 1998 the 33 had grown up into a muscular 1.7 Green Cloverleaf version, the swathes of dull plastic swept away and replaced with colour co-ordinated front and rear valances and sill (rocker panel) extensions.

This is what the 33 dashboard looked like in 1983. There's a wood-rim wheel, with the key dials slightly hidden by the rim, and a more prominent centre panel housing the ventilation and auxiliary switchgear. There's plenty of storage space for oddments in the various cubby-holes, and although it comes over as a bit fussy, there's nothing to really offend.

ticipated the angular uplift that would characterise the Alfa 75 of 1985, and was even present at the C-pillar a decade later in the sleeker 155 model. Typically, the appearance of all Alfas of the period was to an extent marred by acres of dark rubberised plastic trim, which only worked if the body colour

This cabin interior was available across the whole range in 1988, and, frankly, it looks more Golf Club than race track. The Pomigliano-built cars' interiors always seemed to lack the solidity of those manufactured at Arese - a contemporary 75 will feel as if it's a bit better screwed together, as well as having slightly better materials.

Part of the Pomigliano test track in the mid-80s, showing 33s being evaluated on a variety of surfaces, like the humpback bridge in the foreground and banked bend at the rear. Like the 'Sud, the 33 was at its best on fast backroads.

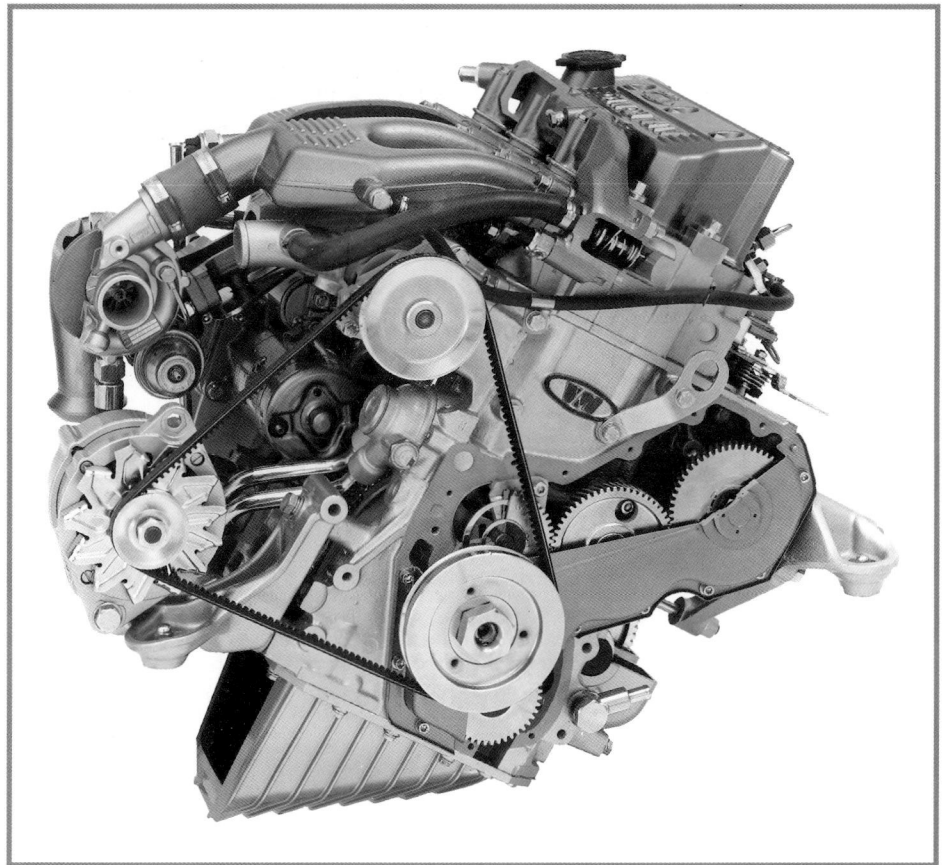

was black. The 33's front suspension layout was different in so far as the Alfasud's brakes were inboard whilst the 33's were outboard, but it was still composed of MacPherson struts, longitudinal arms and coil springs. At the rear was a rigid tubular axle, located by a Panhard rod, with coil springs over damper units and longitudinal trailing arms. The 33 was a heavier car than the Alfasud, which perhaps made it a bit less agile, but performance was adequate and it was fun to drive in a rural environment, coming into its own on faster back roads where flat handling characteristics and good turn-in made for fast progress. Later models such as the Permanent 4 saloon and 1.7ie 16-valve Green Cloverleaf were towering performers in the back road stakes.

It was clear from the outset that the new model needed a bigger engine to keep up with rising standards of performance throughout the industry, and the smallest engine in the range

Powerplant of the 33 1.8 TD was this compact three-cylinder, VM-produced, 1779cc HR 392 SHT unit. It featured a contra-rotating balancer shaft and KKK turbocharger with air-to-air intercooler, and developed 74bhp at 4000rpm.

Designed and built by Pininfarina, the 33 Sport Wagon was always going to be more interesting visually than the basic saloon. Alongside the 1.7 Quadrifoglio Verde, 4x4 and fuel-injected versions was the 1.8 Turbodiesel, which incorporated the virtues of the sports estate with oil-burning economy.

was reserved for the Arna. Thus, at its launch in 1983, the Alfa 33 was powered by the flat-four boxer engine in 1.3- and 1.5-litre formats. The performance figures were similar, propelling it to a 100mph maximum and doing 0-60mph in around 11 seconds. Two of the 1351cc cars were fitted with two twin-choke carbs, and on the single carburettor version the pancake air cleaner was embossed with the cloverleaf symbol, while the capacious windscreen washer fluid vessel was located on the right-hand inner wheelarch. Trim levels differed, so that the 1.3S got electric front windows and a split folding rear seat. This arrangement allowed at least four different permutations of load and personnel to be carried in the rear of the car.

Best equipped of the original 33s was the Gold Cloverleaf - or *Quadrifoglio Oro* - which had the benefit of headlight wash-wipe, as well as the dubious assets of brown bumpers and acres of black PVC on its lower quarters. It could also be specified with four-wheel drive, incorporating a rear transmission system made by Ototrasm of Bari in southern Italy, and a two-section driveshaft sourced from Bierfield. The rear diff was adapted from a standard casing, and while the trailing arm suspension layout was retained, the Panhard rod and coil-over damper units were much more substantial items. The lever for engaging four-wheel drive was just to the left and ahead of the main gearshift

The first facelift occurred in 1988, at which point the car was known simply as the 33, rather than Alfa 33.

The grille was changed to include a chrome horizontal bar, while the seats were upholstered in a check fabric.

The sporting version of the 33 was the 105bhp 1.5-litre Green Cloverleaf introduced in 1984, available in red or

light metallic grey only and marked out by a surfeit of matt black sheathing, bronze tinted glass and a metallic grey side stripe. As the first series 33s were phased out in 1986, the second generation was ushered in. Cosmetic

This is the 114bhp fuel-injected rasping boxer unit fitted in the 33 1.7 IE and Quadrifoglio Verde *models from 1988, when carburettors were finally abandoned. The unit could run on either leaded or unleaded fuel, and in 1989 the catalysed Europa version became available. By 1990 the four-cam 16-valve engine was in production, which seemed a far cry from the original Alfasud's motor.*

117

The Alfa 33 1.5 Giardinetta 4x4 was an interesting exercise in that it combined sports motoring with go-anywhere practicality. By equipping the Giardinetta with four-wheel drive in 1984, Alfa Romeo was very much in the vanguard of evolved traction exemplified by sporting cars like the Audi Quattro and the county set's Subaru Legacy estate.

changes were limited to a new grille, white perspex rear light clusters and 14-hole wheel trims. There were no fewer than nine variations on the theme by now, including the 118bhp 1.7 Green Cloverleaf, the 1.5 ti and 1.5 4x4, as well as an oil burner, the 33 1.8 Turbodiesel. The latter was a rather different proposition to the old Giulia Nuova, however, producing a more sparkling 83bhp. The motor was sourced from Stabilimenti Meccanici VM, and was an interesting 1.8-litre three-cylinder unit, equipped with a turbocharger and air-to-air intercooler, as well as contra-rotating balancer shaft to dampen torsional vibrations. It's doubtful that many in the UK were even aware of its existence, let alone got to drive one, even though it was available from 1986 to 1994.

The 905-series Alfa 33 Giardinetta and subsequent 907-series Sport Wagon estates were also fitted with this engine, which perhaps befitted their more functional rôles better. Hallmark of the 33TD was a black plastic visor that formed a canopy over the rear window. Alfa Romeo also experimented with a high performance diesel before the 33 was launched in 1983. It was fitted with a Comprex supercharger, which took it from 0-60mph in 12 seconds and gave a top speed of 108mph. Economy was an impressive 47mpg.

By June 1988 the 33 was available with a fuel injected version of the boxer engine, with capacity now up to 1.7-litres, in which format it was fitted with hydraulic valve lifters. The range included the base model 1.3S, rising to

the 1.5 TI, the 1.5 4x4 and 1.7 IE, the 1.7 Green Cloverleaf and the 1.8TD. In keeping with the trends, the 1.7 Green Cloverleaf was equipped with a full complement of race-rep add-ons, including a front spoiler, minimal side skirts and a spoiler on the vestigial boot lid. These body kits only work if the car is lowered, otherwise, to my mind, the whole ensemble·looks naff. The full Veloce body kit that could be specified included a deeper front air dam with a splitter below, and less half-hearted side skirts.

Real estates

The 33 design lent itself to conversion into an estate car format a little better than had the Alfasud and, frankly, I prefer the overall appearance of the Giardinetta and Sportwagon to the standard saloon. It instantly becomes a more interesting design, plus it has the kudos of being built by an outside specialist. Because unlike the Pomigliano-built 33 saloon, the station wagon shells were designed - or rather adapted - and built by Pininfarina at the Grugliasco factory and then finished off at the 33 car plant. The Pininfarina badge adorns the Giardinetta's C-pillar to endorse the styling house's handiwork.

The Alfa 33 1.5 Giardinetta harked back to the Alfasud Giardinetta and, indeed, maintained a tradition of specially-made Alfa estates that included the Colli-bodied Giulietta and Giulia

models of the '50s and '60s. First up was the 1490cc 33 Giardinetta 4x4 station wagon announced at the Geneva show in 1984. It became available as a conventional front-wheel drive model the same year.

From a practical point of view (which is, after all, the purpose of an estate), the 33 Giardinetta provided a fair amount of carrying capacity, especially with the rear seats folded down. However, this was slightly compromised by its 4x4 layout. The rear transmission meant the cargo deck was not as low or as deep as it might have been with a front-drive only layout. Additionally, the rear hatch did not extend below the number plate, so access to the interior was only so-so. In my personal experience you could get a bike in, but a washing machine, well, that had to be carried with the tail-gate open as it wouldn't fit over the rear panel.

In 1988, the Giardinetta in 4x4 and front-drive versions was phased out, to be replaced by the 33 Sport Wagon. This was quite an apt name in view of its excellent performance, especially the fuel injected 1.7 Green Cloverleaf model that could top 200kph. In four-wheel drive format, the Sport Wagon with the 1.7 engine was designated the IE 4x4 Europa, and latterly in 1994, its final year of production, it became the 33 1.7 IE 16V Permanent 4 Quadrifoglio Verde S. Bit of a mouthful, that.

The tailgate of the Sport Wagon ends above the light clusters, which means there's a significant lip, over which you have to lift whatever it is you're trying to stow on board. Supermarket shopping bags or school satchels aren't a problem, but if it's something like a washing machine for which you need a flat loading bay, you're in trouble.

This shot of the 33 1.7 Quadrifoglio Verde makes Winter in Italy seem positively inviting. It's the final and most highly evolved evolution of the model, in 137bhp, 16-valve Permanent 4 format, which featured power steering, ABS, smart Recaro style chequered seats and an intelligent four-wheel drive system that apportioned traction according to road condition (which was relayed by sensors).

Anyway, the 33 Sport Wagon was also available as a front-drive model, which had the advantage of a lower boot floor because the rear driveline was absent, and, of course, this allowed more luggage to be carried, but it was nevertheless still handicapped by not having a flat access point.

Also announced in 1988 was the 33 1.3 Sport Wagon S, powered by the 1351cc engine. In 1990 there were a couple of versions running with this smaller engine; one a front-driver and the other, logically, a 4x4. Alfa made a total of 42,644 Sport Wagons, and along with the rest of the Alfa 33 range, production drew to a close in 1994.

My experience of these cars is twofold. When we lived in deepest Herefordshire the local Alfa dealer

119

Amongst the evergrowing list of UK Alfa Romeo specialists is Southern Alfa, based at Southampton, which offers new and used parts as well as sales and servicing. This is the company's promotional vehicle, a late model 1.7 Green Cloverleaf.

used to lend out his personal 33 Sport Wagon 1.7 whenever our 75 was in for something and, on the relatively deserted back roads of the Welsh marches I'm sure there was nothing quicker than the 33. Then a little later, having relocated to north Norfolk, my wife, Laura, had a 33 Sportwagon 1.7 with the Veloce body kit, and it proved an extremely rapid vehicle for making cross-country runs, being more nimble in narrow lanes than a bigger car. It did duty as her commuter vehicle as well as serving on the lengthy school runs, and it was here that its deficiencies became apparent. They really centred on the inadequacy of padding in the seats, indifferent quality of fixtures and fittings within the cabin, and the downright discomfort of the driving position for someone over six feet tall. Why have one then, you might well ask? Well, to start with it was my wife's car, and because, like many Alfas, the total package compensates for individual shortcomings. It's also more affordable and more fun than, say, a BMW 3-series estate.

Top dog

By far the most interesting derivative of the 33 was the 33 1.7 IE 16V Quadrifoglio Verde introduced in 1990. Its 1712cc boxer unit was fed by Bosch Motronic ML 4.1 electronic fuel injection, and now it used two camshafts per bank, producing 137bhp at 6500rpm and capable of a top speed of 208kph (128mph). Fuel consumption was a not unreasonable 29mpg.

Alongside the Green Cloverleaf were marketed the catalysed Europa model and the 4x4 and Sport Wagon. The wheelbase was lengthened slightly from 2455cm to 2475cm, and suspension pick-up points were subtly altered. Along with the rest of the range, a major facelift resulted in a more sloping attitude to the nose, a big air intake below the bumper and deeper front and rear valances that incorporated the front apron, with a bigger race-derived spoiler on the bootlid.

Trim and cabin furniture - like door pulls and armrests - was also improved to a level commensurate with the car's other attributes. Supportive Recaro-style seats sported chequered cloth upholstery. A rationalised dashboard completed the aesthetic improvements. The instrument binnacle no longer moved up and down with the steering wheel adjustment, as had earlier been the case, but the pedals were still too close together for anything over size nine shoes. Equally, the relationship between the pedals and the steering wheel to the seat remained an aggravation for taller motorists. But, at last, the car had matured, down to

the full-width rear light cluster panel which imitated that of the big 164.

In four-wheel drive layout, the model was known as the 33 1.7 IE 16V Permanent 4, which inherited its four-wheel drive set-up from the Giardinetta. As the name suggests, the fundamental difference now was that the system was extended from part-time to permanent mode. The sophisticated transmission system featured a central Ferguson-type viscous coupling, allowing a front-to-rear torque split dictated by prevailing driving conditions and the nature of the road surface. This model ran with a unique pattern of five-spoke alloy wheels. From a practical standpoint, the torque-steer that had been the bane of the powerful 1.7-litre 33 models was now subdued and, complemented by speed-sensitive power steering, the 33 was more surefooted than ever. In competition, the 33 was inevitably a front-runner, and, like the 'Sud, 33s still dominated the Alfa Romeo Owners' Club race series during the late 1990s in classes E and F, and were pretty much to the fore in classes A to D as well. In the intensely competitive arena of circuit racing, such consistency goes to show what an excellent package the 33 and its derivatives evolved into.

In AROC racing circles the 33 is a very effective weapon, as frequently demonstrated by drivers like Graham Heels, whose Class F car is pictured in the Brands Hatch paddock.

As is the case with all Alfas, the 33 can be raced in different classes, according to engine capacity and extent of modifications. This is Andy Curtis' more highly evolved version at Mallory Park, showing front splitter and outsize rear wing.

Alfa Romeo Berlinas

10
TOTALLY
UP-FRONT

In the public eye, the 164 was the model that marked the turnaround of Alfa Romeo fortunes and restored its reputation as a producer of top-class sporting machines. Pininfarina styling cues included the arrowhead bonnet incorporating the Alfa shield, which provided the inspiration for a new generation of cars that would emerge during the 1990s.

At the time of the stock market battle to win control of Alfa Romeo I was a research associate at the Motor Industry Research Unit, monitoring events in the automotive industry, compiling data and relaying our forecasts to clients. We were, therefore, keenly interested in the to-ing and fro-ing of fortunes as the power struggle raged between the titans - the Ford Motor Company *versus* Fiat - to see which would buy Alfa Romeo from the state-owned IRI.

There was a protectionist element about it, to be sure, as the Italian establishment was on the defensive against the might of Dearborn, clearly seen to be muscling in on a treasured Italian icon. Remember, the Americans had but 20 years previously shown the mettle of their money by trouncing Ferrari at Le Mans. So the corporate giant Fiat, which has a finger in just about every Italian pie, was not about to let the revered Alfa marque slip through its fingers; it already owned Ferrari and Lancia, after all. The end came suddenly, Ford capitulated and

Alfa Romeo became a wholly owned subsidiary of Fiat in November 1986. Ford, licking its wounds, consoled itself by hunting down Aston Martin and Jaguar.

The company's new identity was Alfa Lancia Industriale SpA, and its new chairman was Vittorio Ghidella, although he was quickly succeeded by Piero Fusaro. From the outset, it looked as though Alfa's fortunes had turned the corner. There was a new confidence in the wind as the big new front-drive 164 executive car was launched at the 1987 Frankfurt Show. Clearly, it had been on the stocks well before the takeover battle was fought, and it went under the contrived code name of Alberto. This apparently stood for Al for Alfa, Ber for Berlina and To for Torino, in deference to its styling origins in Pininfarina's design studio at Grugliasco rather than Fiat's Turin HQ.

Whereas previous Alfa Romeo Berlinas were the result of collaboration with Bertone, Alfa turned to Pininfarina for inspiration for the 164's

In the body assembly plant at Arese, Alfa 164 bodyshells go down the line, receiving attention from welders prior to the fitment of doors.

three-box shape, and perhaps it was more refined for that. Highlights of the design included the characteristic slash that extended the whole length of the car from stem to stern, coinciding with the top of the head and rear light clusters. Also, the Alfa arrowhead shield shape heralded the styling cue for a whole new generation of Alfas, and was incorporated into the bonnet front, intruding into the bumper and flanked by air intakes that recalled - distantly - the arrangement of 1950s models. And the shallow band of rear lights and indicators that traversed the back of the car was a very neat touch. A Pininfarina badge was placed just to the rear of the front wheelarches.

In fact, the 164's gestation was overseen by the head of Alfa's styling department, Ermano Cressoni, and with Fiat having a vested interest in the product after it went on sale, quality control was strictly observed. The 164 was built at Arese, and most of the steel for the monocoque shell and the exterior panels was treated for protection against corrosion. There had been heavy investment in new manufacturing techniques - CAD and CAM - including extensive use of robots and offline assembly of key components. This ensured that the 164 was better made than any previous Alfa, and build quality was evident in the heavy clunk when you shut the doors. The anticorrosion treatment included galvanising at least 60% of the bodyshell, administering a PVC coating underneath and wax injecting vulnerable box-sections.

As I've already mentioned, the 164

One of the first models available in the 164 line-up was the 2.0i Twin Spark, producing a reasonable 145bhp at 5800rpm from its transverse-mounted twin-cam, which was adequate for this relatively large car, but somehow never quite matching its perceived status.

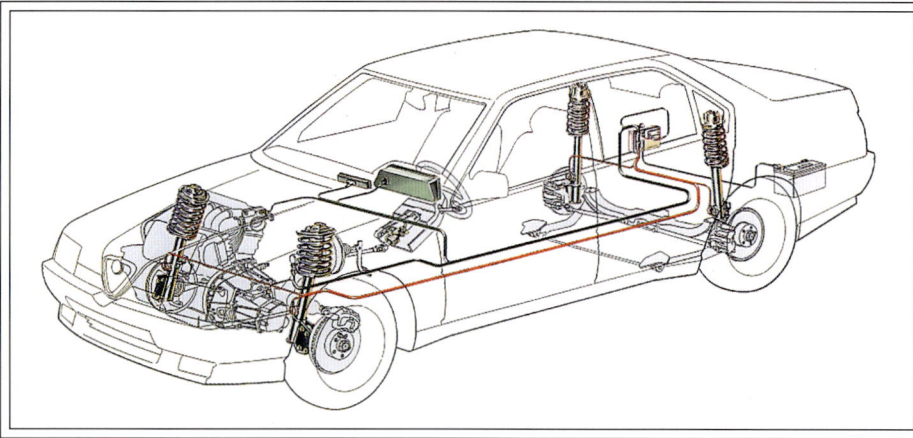

In the 164 Cloverleaf the suspension settings were automatically varied according to vehicle speed.

Middle picture: Diagram of the Motronic injection and ignition system.

Bottom picture: The 164 Cloverleaf's Recaro seat adjustment controls.

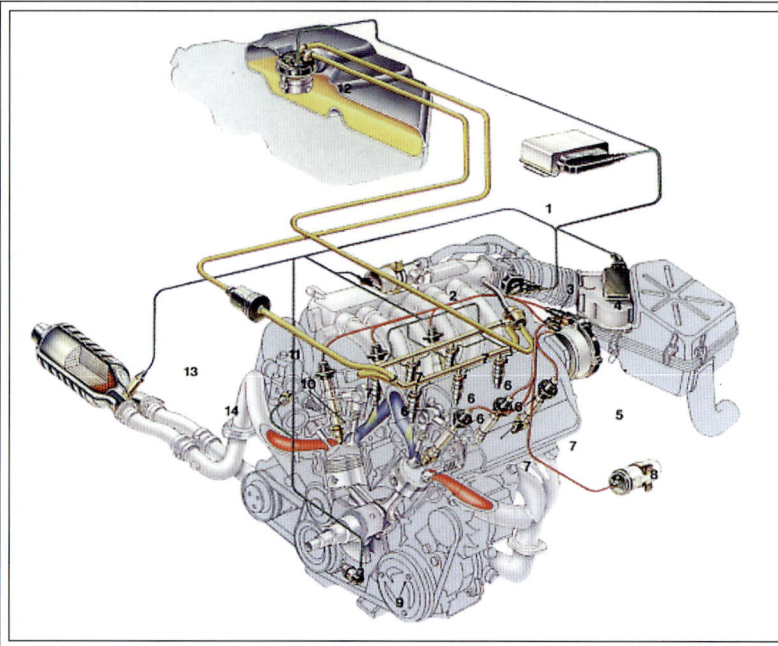

1) Motronic electronic control-box
2) Constant minimum actuator
3) Switch for the min. and max. opening of the butterfly valve
4) Air flow gauge
5) Distributor
6) Electro-injectors
7) Spark plugs
8) Coil
9) Sensor for rev. counter and drive shaft
10) Engine coolant temperature sensor
11) Fuel filter
12) Fuel pump
13) Exhaust system with catalytic convertor
14) Lambda sensor

shared the same platform as the other Type 4 cars from Lancia, Fiat and Saab, so it was always destined to be front-wheel drive, a serious departure from the traditional Alfa drivetrain layout. Models available from the outset were the 2.0i Twin Spark, the 3.0i V6 and the 2.5 turbodiesel. The 164 was hailed as a quantum leap in the international motoring media, by virtue of its elegant lines, panel fit and overall build quality. It was also seen as a mainstream car, rather than something that appealed only to aficionados. Unless you're an Alfa fan, or simply have your eye in, it's possible to dismiss cars that don't instantly ap-

peal or are flawed by the kind of quirks that afflicted the 75. So it was likely that many of the cheerers hadn't previously experienced the wonders of the Twin Spark or 3.0-litre V6 engines, which undoubtedly helped the 164's case. Now, though, the transverse mounted engines delivered their power through the front wheels, and torque steer was the inevitable result, especially in the case of the V6.

Suspension was by MacPherson struts, coil springs and dampers, plus an anti-roll bar at the front, with power steering and ventilated disc brakes allied to a dual-circuit servo-assisted Bosch ABS. At the rear was a MacPherson strut arrangement as well, featuring two transverse arms and a longitudinal arm, together with coil springs and damper units and an anti-roll bar. The 164 was also available with self-levelling rear suspension that raised the rear of the car when heavily laden, based on sensors in the dampers that

Flying the flag for the turbo brigade was the 164 2.0i Turbo, pictured here at the Carlton in Cannes. It was powered by the same engine as used in the turbocharged versions of the Lancia Thema and Fiat Croma, and useful for those who wanted to duck under the 2.0-litre capacity tax threshold and still have a fast car.

activated an hydraulic system and jacked up the car by pressurising the spring and damper units.

The transverse orientation allowed the pair of distributors serving the

Twin Spark engine to be keyed to either end of one of the two camshafts. A Bosch Motronic integrated fuel injection system was fitted, with a twin ignition and static advance system incorporating an electro-hydraulic variable valve timing device. Engines were located on the front subframe via hydraulic engine mounts, which minimised the transfer of mechanical vibrations into the cabin. The V6-engined car was introduced at the same time as the Twin Spark, and if you were in the market for a refined, high performance saloon, you couldn't do much better, provided, of course, that the 'new' front-drive factor didn't faze you. With a wheelbase of 2660cm and overall length of 4556cm, the 164 was some way bigger than the 75 (2510cm and 4330cm respectively), taking Alfa Romeo into a straight fight with executive cars like the BMW 5-

This monster is the Alfa 164 Procar, built in 1988 as a potential contender in the FISA Procar championship for silhouette specials that never got off the ground. Its plastic body faithfully replicated the production model, with the addition of a rear wing, but it was powered by a full-race, Alfa Corse-built V10 engine that was mounted amidships. Amusingly, it was badged as a Twin Spark.

Viewed head-on, the 164 3.0i V6 Green Cloverleaf was a bit of a bruiser, and the performance from its 228bhp 24-valve 3.0-litre V6 engine measured up to the image it projected, with a potential top speed of 240kph (145mph).

The 164 was a big car to go racing with, but at least it got you noticed, partly because of its rarity in competition, and it had a substantial physical presence on the circuit. Here is Jane Cheffings rushing down Paddock Hill at Brands Hatch during an AROC race meeting in 1995.

series. In this kind of company, buying an Alfa or a BMW was going to be a contest between the head and the heart.

The 164's interior was relatively plain and simple, although not up to German standards. It was functional rather than lavish, although managing to be both spacious and comfortable at the same time. The seats were electronically fully adjustable in most directions, with decent headroom in the rear for all except the tallest passengers. Principal gauges were housed in a hangar-shaped binnacle ahead of the steering wheel, flanked in the centre by a daunting array of warning lights and flush-fitting switches - rather like the self-service ticket displays at London underground stations. The window lifter switches and seat actuators lay in another panel alongside the handbrake to the rear of the gearlever. The ventilation system was also done in a monumental way, looking like a miniature apartment block by someone like Richard Rogers (or Norman Foster). By 1992 instrumentation had been slightly revised and a four-spoke 'Porsche'-type steering wheel, with air bag, fitted.

In 1989, the catalysed 164 2.0i

Left - While the race-derived body kit gives the 164 3.0i V6 Green Cloverleaf a mean and hunky look, the vast expanse of matt black plastic also gives the impression that it's a car of two parts. Whilst, granted, a car finished in one body colour might have looked dumpy, a two-tone scheme of some sort might have given a more harmonious result.

Twin Spark Europa and 3.0i V6 Europa were introduced, both of which suffered from a slightly reduced power output; both the cat and non-cat Twin Spark models were fitted with a new gearbox the following year. Also in 1989 the V6 model became available with a sophisticated 4-speed ZF automatic transmission. In practice this worked by means of the torque converter to change first and second gears, with only 60% assistance on third, while fourth gear was entirely manual, by mechanical transmission.

A year earlier, Alfa had taken advantage of the symbiosis between itself and the Fiat family and borrowed the cast-iron block, belt-driven 2.0-litre four from the Croma and Thema parts bin. This was a strong engine which, when fitted with a Garrett T3/50 turbo, provided the basis for a turbocharged Alfa model, the 171bhp 164 2.0i turbo. Both sister companies offered similar forced-induction Type-4 models (as did Saab), so it wasn't a move that required any deep consideration. However, it was the first time that a petrol (gasoline)-engined Alfa was powered by anything other than an engine of the company's own design and manufacture, and devotees remained unconvinced. They may have been right, as only 12,440 units were produced in 1988.

More to the point was the 164 2.0i V6 Turbo, a model launched at Geneva in March 1991. The intention was to provide an Alfa with the sonorous benefits of the V6 engine, whilst at the same time ducking below the Italian 2000cc tax break without losing anything in the way of performance. The 1996cc V6 was based on the layout of the 12-valve, 2.5- and 3.0-litre versions but with reduced bore and stroke (80mm x 66.2mm). It was equipped with the Garrett T3/50 turbo and air-to-air intercooler, with power output quoted at 210bhp. It was sometimes known as the TB, which stood for Turbo Benzina, and in 1992 it received the styling upgrades of the rest of the 164 line-up. Between 1991 and 1996 (when the model was discontinued) Alfa made 109,895 units of this version, so it was hardly a small run special edition.

A 164 to queue for

A more radical version came out in 1994, having been under development since 1992, and this was the 164 3.0i V6 24V Q4. A bit of a mouthful, but an interesting combination of technical advancements nevertheless. It featured the 24-valve version of the well-proven V6, with twin-overhead cams per bank, and a four-wheel drive system incorporating an electronic Viscomatic central coupling. This silicon-filled device adjusted the application of torque through variations in the volume of silicon (by means of a piston) which, when compressed, released maximum torque. The result was an even more sure-footed machine with virtual go-anywhere potential - bearing in mind its low ground clearance - and it was the flagship model in the range.

The conventional 3.0i Quadrifoglio Verde ran with the 24-valve engine, and was treated to fresh bumpers front and rear that extended downwards into a more prominent air dam at the front and skirt at the rear. Matching side skirts were fitted, along with unusual five-spoke alloy wheels, and this model looked like a regular touring car racer. The downside of all that black plastic was that it had the visual effect of splitting the car in two below the waistline, so that a dark colour - black, dark grey or blue - were the best means of camouflaging it. As with all 164s, driving lights were inset into the front valence, where there was also a slatted air intake.

By now, Alfa Romeo had acknowledged that there was indeed a problem with torque steer on early 164s. The issue was addressed in the Cloverleaf model whereby the V6 engine was mounted lower down in the engine bay, thus permitting the driveshafts to run at a more horizontal attitude. The front suspension was revised to include an electronically controlled, semi-reactive variable damping system, allowing the driver to select either 'sport' mode for a hard setting, or 'auto', automatically adjusting the suspension according to prevailing road conditions.

Silhouette series

This Green Cloverleaf model called to mind the stillborn Alfa 164 Procar project. Actually, it was the race series and not the car that didn't get off the ground as Alfa Romeo faithfully did its bit when it looked as though the Procar Championship would take off. There have been 'silhouette' race series since the 1960s, the idea being that out-and-out racing cars masqueraded as their regular road-going counterparts, clad in lightweight bodies replicating the specified model. In fact, Procar wasn't so far removed from the Class 1 ITC (International Touring Car Championship) instigated by the FIA in 1996.

Anyway, the 164 Procar was built by Brabham to run in the Silhouette series mooted by FISA for 1989, and was powered by a normally aspirated Alfa Romeo V10 racing engine mounted amidships like an F1 car or sports/GT prototype. Curiously, the Alfa 164 Procar was actually badged as a Twin Spark, either as a joke or to provide publicity for that model. In any case, the V10 engine had only a single spark plug per cylinder. The demountable plastic bodyshell was the inspiration for the production model Green Cloverleaf, but with a rear aerofoil and huge slick tyres half hidden under the wheelarches distancing it from road-going reality. Unfortunately, the 164 Procar performed little more than shakedown tests in the hands of Riccardo Patrese, while the Alfa Corse-built 3.5-litre, 72-degree V10 was a potential powerplant for an F1 Brabham, but nothing came of that either.

The 164 was never viable as a competition car, being front-drive and far larger than the 33, and the 75, for that matter. But in the British AROC series a 164 was gamely campaigned for a number of seasons by Ron Davidson, Jane Cheffings and Martin Parsons, albeit with no really notable achievements.

From 1990, economy-minded Continental buyers could opt for the 164 2.5 Turbodiesel, which, like the Lancia Thema, was fitted with the four-cylinder, 2499cc VM Turbodiesel engine. The Alfa version of this engine differed from the Lancia's in having a somewhat old fashioned camshaft location - the gear-driven shaft was housed in the crankcase and operated the valves by pushrods and rockers instead of an overhead cam and pulleys. Despite giving away a couple of horsepower to the Lancia turbodiesel, the 164's engine was the torquier proposition, running up to over 200kph. The turbo was an intercooled KKK unit, and the Alfa was the fastest Italian oil-burner on the road in the early 1990s.

In 1992, the standard and Lusso V6 models were joined by a more highly specced Super version, which included a number of slight trim revisions inside and out, most notably to bumpers, valances, front and rear spoilers and side mouldings, all of which were made less acute. The Twin Spark Super came out the following year, while the Turbodiesel received the Super treatment as well, and its power output in Super form rose to 125bhp at 4200rpm. Alfa Romeo called time for the 164 in 1997, at which point 273,407 units of all versions had been produced.

There's no doubt that the model did an excellent job in re-establishing the company as a credible player in the mainstream market.

11
BACK
ON TRACK

The 155 was the second new model to appear under Fiat's ownership of Alfa, and, like the 164, it was front-wheel drive with transverse-mounted engine and transmission packages.

I'm on home ground again here, as the current family hack in the Tipler garage is a 155. It's a 2.5-litre V6 Green Cloverleaf model, dating from 1995, and was acquired to replace the 75 - which was apt since that was the model's original purpose in the grand scheme of Alfa Romeo production. When it became clear that our 75's days were numbered, I thought I'd better have a test drive in a 3-series BMW just to make sure that the rear-drive option might still be a possibility. As I touched the ton to check its acceleration during the test drive, I hadn't reckoned with the following presence, albeit lagging some way behind, of the unmarked Mondeo. Only when it caught up at some traffic lights did I notice the blue shirts of its occupants, and that innocent little venture cost me £600 including costs, plus six penalty points. I therefore harbour a strong, if irrational, grudge against BMWs, and found any number of reasons to support the purchase of the 155.

To a great extent, the 155 was

Manufactured in the Arese factory, the 155 bodyshells were created on fully robotised production lines. Sunroof apertures are already present.

129

built by robots on the Arese production lines, and over 70% of the body panels were galvanised. Prototypes underwent some four million miles of test driving during pre-production development. Like the Lancia Dedra, the 155 was based on the communal Fiat Tempra platform, and was designated the *tipo 167*. It was introduced at Barcelona in January 1992 in 1.8i and 2.0i Twin Spark format, as well as a four-wheel drive model known as the 2.0i turbo 16V Q4, and a 2.5i V6. From 1993 there were also the tax-friendly 1.7i Twin Spark and 1.9 and 2.5 turbodiesel versions. Although broadly similar in size to the 75, it was 110mm longer overall, with a greater overhang at the front and a 30mm longer wheelbase. It also differed radically in being front-wheel drive like the 164, and, although build quality was far better than the 75's, there were still some extraordinarily inconsistent panel gaps on the new model.

At the outset, styling of the 155 was a cross between the 164 and 33 with a fresh set of detailing, yet it still

The original 155 of 1992 had only slightly flared front wheelarches, and rear ones that retained the line of the rest of the body's side panels. That would change with the so-called wide body revamp of 1995, although features like the slim headlights and driving lamps in the front valance stayed put.

The bluff, high tail of the 155 contained light clusters with slim indicators at the top, and their angled corners created an elongated diamond-shaped space for the numberplate. An enthusiastic owner has added an extravagant rear wing.

retained certain elements of the old 75. In short, there was a strong family resemblance between the outgoing ranges and the contemporary one, which, at the same time, managed to progress the look of the Alfa saloon a stage further. The 155 was the product of computer-aided design, which may have been instrumental in creating a more harmonious body shape than had the 75. It was fundamentally a product of the wedge school of design, with a low and compact frontal area rising to a particularly high boot that made the car problematic to park for some. It wasn't short of interesting details however, including the V-shaped moulding on the bonnet that began at the Alfa shield. This item was inset into the top of the bumper, with narrow air intake bands on each side and equally narrow headlight and indicator clusters. As with the 164, driving lights were set in the front valance; in practice these are virtually useless as they're difficult to adjust, and are generally only useful for annoying roosting birds in the hedgerows. At the rear the area in between the rear light clusters formed an elongated diamond shape.

The 155 suspension consisted of

The 1995 edition of the 155 was given the brand new 1970cc Twin Spark engine, designed by Ing Alessandro Piccione and his team, with a lightweight cast iron block, two spark plugs and four valves-per-cylinder, contra-rotating balancer shafts and variable valve timing.

Powered by what was effectively the same engine as in the Lancia Delta Integrale, the 155 2.0i turbo 16V Q4 was a 186bhp rocketship. With turbo boost and four-wheel drive, it was not only fast but surefooted as well.

MacPherson struts and lower linkages, with coil springs and damper units, mounted on a subframe, and dual circuit ventilated outboard discs. Steering was power-assisted rack and pinion. At the rear was a beam axle with trailing arms, coil springs and dampers and an anti-roll bar, with outboard disc brakes here as well. It was all beefy stuff. An adjustable damping system was available as a cost option, with the Sport and Auto settings operated by a dashboard control.

My first meeting with a 155 was in 1993 when I worked as a sub-editor on the motoring newspaper *CarWeek*, and had access to a 1.8 press car for the weekend. Its build quality and sturdiness were immediately apparent, and although power delivery fell some way short of inspirational, I do remember it was great fun on deserted roundabouts. It was clearly also a competent package, and served to whet the appetite for a go in a 2.0 Twin Spark.

One version that never impinged on my consciousness was the turbodiesel - which came with 1.9- and 2.5-litre engines that laboured under the contrived Ecodiesel label - because they were never officially imported into the UK. I've nothing against diesel engines - in their place, which is to say in a truck or a boat - but they just aren't sexy in a sports saloon. However, amongst the common herd of Eurodiesel saloons, I'm sure that the 155 was well able to hold its own. The 2.5-litre VM unit was branded the *deinquinato* 96 in 1996 to underline its environmentally friendly credentials. Hmmm ...

Cast-iron case

The major change that took place in 1995 was the introduction of Alfa Romeo's brand new twin-cam engine. Designed under the direction of Ing Alessandro Piccione, this was a lightweight cast-iron block, which, at a stroke, rendered the venerable aluminium twin-cam obsolete. The internals of the new Twin Spark motor remained the same, although contra-rotating balancer shafts were fitted, and the timing gear was operated by a toothed belt instead of the traditional duplex chains. The Bosch Motronic 1.7 fuel injection and engine management system was retained, as was the neat electro-hydraulic adjustable valve timing device.

The four-wheel drive Q4 (which stood for Quadrifoglio 4 x 4) was powered by an Alfa-badged version of the 190bhp 16-valve Garrett T3 turbocharged motor as used in the Lancia Delta Integrale. The front differential was conventional, with a two-piece propshaft split by a Ferguson viscous coupling allied to an epicycloid unit at the centre, and a Torsen limited slip diff at the rear. Sensors indicated road surface conditions and the Ferguson system automatically transferred torque to the appropriate wheels for optimum traction. With full-time four-wheel drive, grip was said to be prodigious, and the Q4 differed from the other 155s in having Weber-Marelli electronic ignition rather than Bosch Motronic, This gave it a top speed of 225kph (135mph), making it the quickest of the 155s. However, it was only available in left-hand drive and gained

a reputation among Alfa specialists as being overly complex and difficult to fix. Like other 155s in the range, it could be specified from 1993 with the Sport Pack options including black five-spoke Speedline alloys, 205/45 ZR 16 tyres and a boot-mounted aerofoil. The Q4 was dropped in 1996.

The 155 2.5i V6 model used the 60-degree 166bhp 12-valve 2492cc engine - mounted transversely now - and was capable of 215kph (130mph). The V6 delivered power and torque smoothly, and its chromed inlet manifolding was a work of art compared with the plenum chamber that concealed the plumbing on the workmanlike 75. Engine management was a fully mapped Bosch M1.7 system, and a totally new compact alternator and micrometric tensioning system for the drivebelt was also fitted. The gearshift mechanism of the five-speed manual box was by Bowden cables on the V6 and rods on the Twin Spark, and absolutely light years more positive than the transaxle arrangement of the 75's old Alfetta derived system. Having said that, it was advisable in Wintertime not to hurry the second gear until the engine had warmed up. There was also power steering that decreased in effect the faster you went, but was well up to the job of parking in tight spaces in town. Under these circumstances, a malfunctioning power steering belt could result in a banshee-like wail on full lock. From 1993, all cars were fitted with catalytic converters - for legal reasons as much as anything - and passive safety features, including side impact bars and air

bags, were introduced throughout the range.

Quick rack

Although the V6 model never received the 24-valve quad-cam engine, the real revelation came in 1995 when the 155 got the 'quick' rack, ushered in with the new Spider and GTV, requiring only 2.1 turns lock to lock, which sharpened up the turn-in significantly. The engine was set lower in the chassis to improve weight distribution, and it also got wider track suspension. It had a diagonally split brake circuit, and a proportioning valve on the rear axle ensured that the braking effect between front and rear took into account the weight and distribution of the car's payload - like, whether there was luggage in the boot and rear passengers *in situ*. To accommodate these suspension mods, the wheelarches were flared accordingly, giving the car a much more purposeful stance altogether. These are often known as 'wide body' cars.

As I mentioned, the car I run is a 155 V6 Green Cloverleaf model. This was a limited edition model built in 1995 which looked virtually identical to the 4x4 Q4, including the 6.5J Speedline fake split-rim wheels, but

differing in so far as this is a 'wide-body' car with the exaggerated wheelarches of the later model. There are also hints of its recent Super Touring ancestry in the embryonic splitters that protrude from the front and rear skirts in a similar way to the 164 Green Cloverleaf. As far as I can tell, the only difference to the standard 155 V6 model is its uprated suspension, which my accomplice Richard Drake has aided a bit by fitting shorter springs and lowering the ride height. It has a fairly uncompromising ride, much bumpier than the 156, for instance, but is by no means uncomfortable. When required, it can behave like an overgrown Go-Kart, so superb is its turn-in, and it can be more thrilling than the 75, probably because it's more confidence-inspiring.

The single disadvantage is the torque-steer element, which can be quite alarming if you're trying to overtake in a hurry on an undulating rural backroad. It's all over the place, requiring judicious use of the throttle, and it can be a real wrestling match to keep it out of the scenery. My guess is that's why Alfa kept it at 2.5-litres rather than install the 3.0-litre unit, although that may have been to restrict it to a performance level below

that of the 164. You have to use more revs to keep the performance up, but otherwise it's not much different to 3.0-litre power. The 155's Bosch ABS brakes were a revelation after the hopelessness of the 75's, and these further enhanced one's feeling of confidence in the car. It's now shod with Yokohama A520 tyres, which are excellent and help counteract torque-steer.

And then there's that practical

The model's 1995 facelift included flared wheelarches front and rear, designed to accommodate the wider track suspension. Under the bonnet, the engine was set lower down to improve weight distribution and handling.

The 155 moves in fast circles. Controls are well laid out, including a nicely contoured Momo wheel with airbag, and clearly readable dials ahead of the driver. Bonnet release is below the steering column and window buttons and electric mirror adjustment are above the door pull.

Equipped with ABS, a 'quick' rack and suspension consisting of MacPherson struts at the front and trailing arms at the rear, plus anti-roll bars, coil springs and dampers all round, the 155 was a superb driver's car and capable of good performance on any kind of road.

element again - which you don't get with fashionable coupés - a family of four, plus dog, can travel in considerable comfort. The lack of transmission tunnel makes for a more spacious cabin, with slightly more legroom in the rear than the 75 had. The seat upholstery is simply plain tweed, but more hard-wearing and thus practical for my menagerie than our previous car. There's a full complement of all the stuff you expect in highly-specced cars nowadays, like electric windows, electric sliding and tilt sunroof, with comprehensive instrumentation and face-off radio cassette player and six-speaker stereo system. The principal dials are flanked by oil pressure and temperature indicators, with no fewer than 22 warning lights in a panel on the centre console below the heater and ventilation controls. The heated rear window and rear foglights have fingertip sensitive buttons on the ends of lights and wiper stalks, and there's a boot release button hidden in the glove locker. Driving light switches are on the centre console, along with rear window switches, while front window buttons are in the door armrests. It also comes with rear window blinds to keep out that hot Mediterranean-style sunshine that we get so much of in Norfolk. It has a nice contoured Momo wheel with air bag, and even the driving position is close to being acceptable; there's a lumbar adjustment which you can firm up to give your back more support, although I'm not convinced that it actually makes much difference. You have to be aware of not allowing yourself to shrink into the seat on a long run - I

find it's important to place your backside right back into the seat to start with.

If there's one criticism I have about the cabin it's the demister, which takes forever to clear the windscreen on a

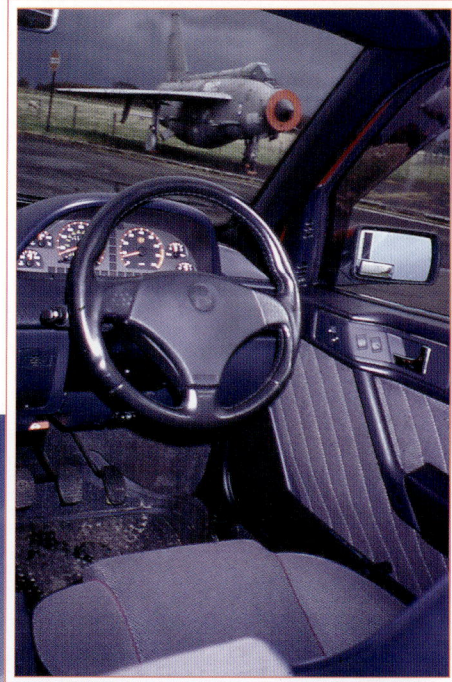

school run during the Winter. The boot is vastly superior to the 75's, thanks in part to the spacesaver spare, and the fully opening bootlid provides a flat loading bay.

With hindsight, maybe I should

During the gestation of the 155, all functions were extensively tested and monitored, like the dashboard functions set up in the Arese test facility, pictured in 1992 just prior to launch. During the build process, the fascia assembly is placed in the body automatically by robot arm.

Boot space in the 155 is generous, especially compared with its pred-ecessor, the 75, and this is partly because a spacesaver spare is carried under the fold-back boot floor. The bootlid is supported by a pair of hydraulic struts.

have saved up a few more Euros and gone to the Continent to get a 156. A 1.8 model 156 bought in Amsterdam would have cost only a few grand more than the 155. However, at the time of writing, the V6 156 is still a long way off my budget, so I reckon the 155 is a good compromise for perhaps half the money, having the same suspension and steering as the 156, if not the 24-valve engine, plus similar carrying ca-pacity and what is still quite an out-standing design. Manufacture of the 155 ceased in 1997, when total pro-duction of the model topped 465,000. Rarest version was - surprise, surprise - the V6, with less than 6000 units made.

When there's not too much money in the kitty, a late model 155 is almost as good a proposition as a new 156. This one's a 2.5i V6 Green Cloverleaf from 1995, which was a limited edition car with uprated suspension. It also ran with the perforated Speedline alloys that imitated the split-rim competition variety.

Superturismo!

With the 155, Alfa Romeo was back on the scene in Touring Car racing, having been absent, in the UK at any rate, for several years. The original 155-based competition car was developed early in 1992 as the GTA and was based on the Q4. A front-drive version was built up to comply with the FIA's Class 2 regulations and tackle the burgeoning Super Touring category that was developing into a series of national championships all over the world. The project was supervised by Alfa Corse managing director Giorgio Pianta, general manager Pierguido Castelli and design engineer Sergio Limone, while the 23-strong squad was managed by

135

Above: The 155 provided the ideal vehicle for Alfa Romeo to return to the International Touring Car racing scene, having kept a low profile outside of Italy for a number of years.

The icing on the cake for Alfa Corse was Gabriele Tarquini's victory in the UK's BTCC in 1994; Gabriele celebrates the win along with team manager Nini Russo and team-mate Giampiero Simoni. The cake looked so good it was a shame to cut it.

former Fiat Abarth and Lancia rally star Nini Russo. They were allocated what seemed at the time a vast budget of £5m with which to contest the British Touring Car Championship, but which, in effect, simply caused other teams and their manufacturers to follow suit.

Drivers for the 1994 BTCC were to be ex-F1 ace Gabriele Tarquini, backed up by the eminently competent Giampiero Simoni. Four 155s were built up - two to be raced while another couple were being refettled in Turin. The shells were seam-welded and stiffened massively by the complex trusses of their roll cages. Suspension consisted of MacPherson struts and gas-pressurised Bilstein dampers complemented by Eibach springs at the front, with fabricated steel trailing arms at the rear. Bolt adjusters on top of the suspension turrets were used to tune

Giampiero Simoni, left, and Gabriele Tarquini pose with one of the 155 TS Silverstone cars they campaigned in the successful bid to win the 1994 BTCC. Gabriele had previously raced 155s with some success in the Italian and Spanish Superturismo series, and his eight victories in the BTCC made him clear winner.

Left - Alfa Corse took advantage of the FIA's Class 2 rules and built a stock of engines based on components sourced from the Fiat corporate parts bin. The 155 TS developed 290bhp at 8500rpm.

The boot space of the competition 155 serves as a logical and convenient site for electrical ancillaries, the carbonfibre-funnelled fuel filler and 70-litre racing tank.

the camber settings. The 155's hydraulic power steering was sourced from the Lancia Delta S4 Group B rally car, and braking was by means of massive Brembo eight-piston calliper ventilated 15in discs at the front, and four-piston callipers at the rear, fitted with carbon metallic pads. Wheels were Speedline-cast MIM, measuring 18in diameter x 8.25in wide, using Michelin racing covers.

The Alfa Corse badge on the cam cover indicated that the cars were powered by a very special engine. In fact, this was something of a hybrid, which perhaps only a company with the resources of Fiat could provide. It was entirely legitimate, of course. The leg-

From this ghosted drawing of the 155 TS Silverstone, you can see how the roll cage is diagonally braced, plus the orientation of the engine and gearbox, brake cooling ducts and exhaust system.

The cockpit of Gabriele Tarquini's 155 is an amalgam of the austere and the high tech. It carries an on-board computer, digital rev-counter, sequential gearshift, control for adjusting damper settings and a plumbed-in fire extinguisher system. More basic equipment consists of a suede Momo wheel and Sparco racing seat with six-point harness. Brake and accelerator pedals almost merge into one.

The body shell of the 155 Super Touring was seam-welded and stiffened substantially by the scaffolding that constituted its comprehensive roll-cage. The safety net across the driver's window allowed fresh air in, and prevented the ingress of debris.

endary tuning concern Abarth was entrusted with its construction, and working within the regulations of TOCA, the organising body, Abarth built up the 155 Super Touring engine utilising components from a variety of Fiat corporate sources. The cast-iron block came from the 164's 2.5-litre VM turbodiesel, the alloy head from the 155 Q4 - which, as we know, was basically an Integrale 16-valve - and then rotated through 180 degrees to improve engine breathing. Power was transmitted via an AP racing clutch, solid steel-billet driveshafts and a limited slip diff.

The engine's internal dimensions were reworked in the course of construction and internal components and ancillaries sourced from a number of Fiat parts bins. The whole transverse-mounted unit was canted backwards at an angle of 27 degrees to compensate for the stresses and weight transference inherent in the front-wheel drivetrain and racing suspension. The 155's ignition and induction system consisted of a TAG 3.8 multi-point sequential and programmable fuel injection system with digital electronic ignition. A single 70-litre fuel cell was housed below floor level in the boot, and filled up by way of a carbon fibre funnel.

Performance figures were predictably high, yet par for the course for a racing saloon: peak torque came at 7000rpm, with 290bhp available at the formula's mandatory 8500rpm limit. Abarth held a stock of no less than 40 similarly-specced engines in order to service all 155s active in vari-ous Super Touring championships around the world.

Within the cabin, the cockpit of a 155 Super Touring car was a purposeful contrast of austerity, structural rigidity and hi-tech laboratory. The bare metal of the trimless interior was criss-crossed with the bars of the complex roll-over cage, and home to a single Sparco racing seat with the characteristic side pieces either side of the driver's head. Instrumentation consisted of just a digital rev counter, minimal switchgear and an on-board computer; controls included the sequential shift for the Hewland six-speed gearbox, a suede-rimmed Momo wheel, drilled pedals and a full six-point racing harness, all rounded off by a plumbed-in fire extinguisher system.

Victorious debut

When the 155s of Tarquini and Simoni walked away with the victory at their first appearance in the 1994 BTCC, there were howls of disapproval about their aerodynamic additions. These consisted of a two-position front splitter and big rear boot spoiler, evolved in the Fiat wind tunnel in order to eliminate lift and generate downforce. Several teams - led by Ford - protested, and the Alfas were temporarily suspended. I went to the BTCC round at Oulton Park immediately after the ban - ostensibly to interview John Cleland and Andy Rouse for *CarWeek* magazine - and the 155s were surrounded in the paddock by extremely long faces. They didn't turn a wheel at Oulton that day, and stayed away the round after that as well.

In order to appease the protestors, the front splitter was retracted and the rear wing extensions omitted. Despite missing a couple of races, Tarquini still went on to clinch the BTCC title. That included a spectacular roll-over at Knockhill, from which Tarquini emerged miraculously unscathed - testament indeed to the strength of the roll-cage. The gallant Simoni ended the 1994 season in fifth place, and by way of commemorating this success, a limited edition car appeared in Alfa showrooms, dubbed the 1.8 Silverstone. This model featured the Sport Pack additions of lowered suspension, black Speedline alloys and rear aerofoil.

Although the 155 continued to be a major force in Super Touring elsewhere in Europe, in the BTCC the other teams had caught up by 1995. Thus, after much hype about the Old Spice-sponsored cars, launched at a windy Snetterton test session, and appointment to the number one seat of Formula 1 star and World Sportscar champ Derek Warwick, with Simoni as running mate, the Prodrive-run team floundered. This was in spite of the best efforts of Prodrive's Dave Benbow to sort out the new wide-track suspension set-up of the new season's dramatic-looking wide-bodied cars. The real dilemma was that the engines were now way down on power, which was probably why Derek, used to controlling prodigious amounts of power through the rear wheels, never quite seemed to get a grip of the 155.

In mid season, Giampiero Simoni was packed off to race in the Spanish

Left: Giampiero Simoni heads the BTCC pack into Redgate corner at Donington Park during the 1994 series. Simoni, then aged 26, raced in Formula Ford in the UK and F3 and F3000, before joining Alfa Corse. He finished the '94 BTCC season in 5th place, having scored one win at Brands Hatch.

Right - Simoni's 155 dives into the Craner curves at Donington, pursued by one of the Renault Lagunas. The car's rear wing and heat protection plate over the rear valance where the exhaust pipe exits are clearly visible.

Left: The Superturismo 155 TS sits squarely on the track, displaying several degrees of negative camber at the front. Four of these cars were built specially to contest the BTCC, whilst others were campaigned in the Italian, French, Spanish and German national series.

Superturismo series, and Gabriele Tarquini was brought back into the BTCC squad in a vain bid to regain lost ground. Unfortunately, neither of the superstars could overcome the 155's deficit, but that's the way it goes some years.

The 155s may have been off the pace in the 1995 BTCC, but elsewhere it was a different and less depressing story. That's to say that Luis Villamil kept the Alfa Corse flag flying by winning the Spanish Superturismo championship that year in a 155 TS. Other contenders in 155 TS or 155 GTAs included Giorgio Francia, Oscar Larrauri, Nicola Larini, Gianni Giudici, Fabrizio Giovanardi, Sandro Nannini, Stefano Modena, Gianluca Roda, Giambattista Busi, Enrico Bertaggia, Gherardo Cazzago, Moreno Soli, Felice

Five years on from its glory days, and the 155 still features in motor sport at club level, as witnessed by this shot in the paddock at Spa Francorchamps in 1999. Note the sophisticated front air dam and vented splitter, plus elevated rear wing.

Tedeschi, Gordon De Adamich, Danilo Mozzi, Gianni Morbidelli, Paulo Delle Piane, Tamara Vidali, Antonio Tamburini and Danny Sullivan, most of whom contested the Italian Superturismo series, and a few ran in the German ADAC D2 series in Alfa 155 Twin Sparks.

Rise and fall of the Big Bangers

In 1995 the sport's governing body, the FIA, announced the inauguration of an International Touring Car Champion-ship (ITC) for highly modified Class 1 cars, based on the successful high-tech German DTM series. In the event, only three manufacturers participated, which was a great pity. They were Opel, running with the Cosworth developed V6 Calibra coupé, Mercedes-Benz with the big C-class model built by specialist AMG, and Alfa Romeo, using the 155 V6 Ti specially built by Alfa Corse.

For this series, the maximum per-mitted capacity was 2.5-litres and six cylinders, which suited Alfa very nicely.

Antonio Tamburini, who was Tamara Vidali's team-mate in the German D2 Series in 1994, poses here with his 155 TS. Antonio came to touring cars after a career in F3 and F3000, and raced a variety of Alfa Romeo sa-loons, including a 75 Turbo in 1991, and a 155 GTA and 155 TSs thereaf-ter. He was 2nd in the Italian Superturismo series in 1994, and 4th in both 1993 and 1995.

Above - Tamara Vidali was Italian ladies saloon car champion three years running from 1990 to 1992, the first year in a VW Golf and then in an Alfa 33 16-valve. In 1993 she was 6th in the Italian Superturismo series in a 155 TS, and 10th the following year.

Centre - Among the Nordauto squad for 1994 in the German Class 2 Supertourenwagen series was Tamara Vidali. She also drove an Audi A4 in the 1995 D1 ADAC Super Tourenwagen Cup, (in which Alfa didn't participate) and was up with the likes of Frank Biela and Hans-Joachim Stuck.

Cars competing in Class 1 and the German DTM series in 1994 and '95 could have bodywork made of any material, so long as it looked like the production model, and wheelarches could be as big as necessary to accommodate the 19in diameter O.Z. wheels.

Both Alfa and Opel went in for four-wheel drive, while Mercedes-Benz retained its rear-drive format. Apart from that, they were comparable, highly evolved designs, which almost always produced close and highly competitive racing from twenty-car grids.

While the Class 2 Super Touring cars - such as those that scooped the honours in the '94 BTCC - bore some resemblance to our everyday shopping trolleys and repmobiles, these Class 1 Touring Cars were out-and-out racing cars. Beneath their lightweight bodyshells lurked special space frame chassis, and, in the case of the 155, the V6 engine was slung low in the engine bay to give it as low a centre of gravity as possible. Turbos were not permitted, but these special racing engines still developed somewhere in the region of 500bhp. Imagine that in your family pride and joy; the regular 155 2.5 V6 manages 165bhp! Furthermore, the four-wheel drive ITC Alfa was equipped with an automatic computerised six-speed gearbox, with three electronically controlled diff-locks.

The 2.5-litre V6 engine of the highly modified Class 1 car was set very low in its spaceframe chassis, and was actually based on the geometry of the Montreal V8 motor, generating a colossal 500bhp. This was within the scope of the regulations, which stipulated a ceiling of 2500cc and maximum of six-cylinders. Turbos were forbidden, but induction and exhaust manifolds, ports, fuel injection and flywheel could all be freely interpreted.

The Alfa 155 V6 pictured here at Donington Park, when the Class 1 circus came to the UK on July 9th 1995, was entered by the Schübel Engineering team and driven by Christian Danner, who finished 12th in Race 1 and 9th in Race 2. Aerodynamic modifications were free below the line of the wheel hubs, so the bodywork featured all manner of skirts, flat bottoms and spoilers designed to generate huge amounts of grip.

Suspension was pure competition set-up, with double wishbones all round, hydraulically adjustable anti-roll bars that were controlled from within the cockpit, ABS brakes with eight-pot callipers, and power steering, with Michelin slick tyres on massive 19in diameter O.Z. racing wheels that tended to dwarf the car! The castellated arrangement of skirts and air dams that comprehensively surrounded the 155's body base endowed it with a certain pyramid-like quality, while the aerofoil mounted at the rear of the bootlid wouldn't have disgraced a Sopwith Camel.

Drivers recruited by the four teams that were running Alfa 155 V6 TIs for the ITC came from a variety of backgrounds. There were former Grand Prix aces like Christian Danner, Stefano Modena, Michele Alboreto, Sandro Nannini and Nicola Larini, with several top class drivers like Giancarlo Fisichella (destined for F1 with Benetton), world rally star Markku Alen, Michael Bartels, Jason Watt, Phillippe Gache, Giampiero Simoni, Pedro Couceiro, Stig Amthor and Gianni Giudici as well. As it turned out, Larini and Modena were probably the most successful Alfa drivers in a series pretty much dominated by Mercedes-Benz. The Alfa Romeo teams for 1995 were Alfa Corse 2, Martini Racing, Euroteam and Schübel Engi-

neering, and for 1996 they were lined up under the Martini, TV Spielfilm and two Alfa Corse works JAS banners.

Races in the ITC series were staged at a variety of venues, ranging from converted street circuits like Helsinki city centre, to purpose-built autodromes such as Hockenheim, and on tight tracks like the old fashioned Avus. One downside of the regulations was that top five success earned the driver a weight penalty, with up to 50kg (112lb) of ballast added. There was even some limited television coverage, from which it was clear that it was a highly entertaining formula - as demonstrated when the circus came to the UK at Donington in 1995 and Silverstone in 1996 - and contact was frequent and occasionally violent. But, despite having demonstrated much promise, the 26-round series collapsed at the end of 1996 in financial and political acrimony amongst manufacturers and organisers.

143

12

SMALL IS BEAUTIFUL

The 145 recaptured the spirit of the original Alfasud, being compact, agile and stylistically characterful. Pictured here is the facelifted 1999 car, which featured reworked side-grilles and colour coded bumpers.

The little gem that emerged in 1994 to reclaim the crown once worn by the Alfasud was called the 145. More than the four-door 33 ever had, this three-door hatch embodied the spirit of the 1970s icon and, what's more, its unusual styling cues made it irresistibly cute.

When it first appeared, the *tipo* 930 145 ran with the same flat-four engine as the 33, which made the difference even more obvious. The four versions available were externally identical, but offered with the 1.3 IE, 1.6 IE and 1.7 16-valve flat-fours, plus a transverse-mounted four-cylinder 1.9 turbodiesel. The 1.6 IE was equipped with hydraulic tappets derived from the old 33's 1.7 engine, plus the MPFI Multec-XM integrated injection and ignition management system, and developed 103bhp at 6000rpm. The 16-

Cars with large areas of black plastic - like the Ford Ka and in this case the early 145 Cloverleaf - tend to look better if the body colour is also black. This shot highlights the rear window arrangement that has an opening side window and glazed rear three-quarter panel, making for good rear visibility. The Cloverleaf motif is sometimes blue on the 145.

Among the points of stylistic interest in the 145 - in this case the 16-valve 1.8 TS model - is the shallow V-shape of the rear window base and the narrow bands of the rear light clusters that encroach into the hatch and the corners of the wings.

valve 1.7 16V was a compact little powerhouse that could deliver a punchier 129bhp at 6500rpm, and could take the 145 to a top speed of 200kph (125mph). The 1.6 version was by far the most numerous in production volume.

I tried one with the entry-level power unit - a 90bhp 1351cc fuel injected model, borrowed from Marshall's, the Peterborough Alfa dealer, because I liked the looks and a trade-in was a possibility. However, its chattering boxer engine note reminded me too much of the 33 Sport Wagon that we still owned. Actually, one area where the 145 was made deliberately quieter was in the use of cable gear linkage, specifically intended to reduce cabin noise and vibrations through the gearlever. The five-speed 'box featured internal selector rods and an in-unit differential, along with new synchromesh on first and second gears. Suspension was by MacPherson struts, coil-over dampers and anti-roll bar at the front, mounted on a separate subframe, with trailing arms, coil springs and short, separate dampers forming a low-set independent arrangement at the rear. In practice, this helped give the car a larger loading bay and lower platform.

Somewhat later, my local dealer, Lindfield, lent me a Green Cloverleaf version with the 2.0-litre twin-cam fitted and that was a real hoot. Not quite the family man's Berlina, but an excellent little car nevertheless. I'm a great believer in putting a big engine in a small car, which is perhaps to do with a liking for the wolf-in-sheep's-cloth-ing scenario, and the 145 in this format certainly fitted the bill.

The 145 was the product of Fiat's *Centro Stile* design house, of which Walter da Silva was the prime mover, based within the company's Turin headquarters. The origins of Centro Stile go back to 1958, when it was set up by Mario Boano, another legendary stylist, and his son Paulo, as Fiat's Central Styling Centre and Design School. Some of this pair's early work included the similarly jewel-like Fiat 850 Spider and Coupé and the worthy-but-dull Fiat 124 Coupé.

Friend Walter made sure there were plenty of styling hooks to win over even the most jaded motoring hack, and back in '94 at the 145's launch, the serious journals were beginning to sit up and take notice. Viewed from straight ahead, the model's radical styling featured the pronounced arrowhead nose - or beak-shaped bonnet, if you prefer a different analogy - flanked by air intakes that blended with the poly-elliptical halogen headlights which emphasised the pointed *facade*. Actually, the crucial Alfa shield played a minimal part in this. Clearly designed as a hatchback, the two-box, wedge-shape 145 had virtually zero overhang behind the rear axle. The rear three-quarter window of the 145 ensured good rearward visibility, while the design cue that really won you over was the shallow-angle 'V' of the rear hatch window's lower edge, which was matched by the shape of the rear light clusters. A further embellishment that makes a practical statement is the dip in the window-line that happens half-way along the top edge of the doors. Like the wide-body 155 seen in 1995, the 145's front wings were splayed outwards in order to both cover the telephone-dial alloys and give a racey appearance.

If the 145 was a stylistically tidy little package, it was also very well made. A brief comparison with its Punto cousin in the adjacent showroom establishes that it is exceptionally robust, and its compact shell conformed with the prevailing safety and collision absorption legislation, particularly in the area of side impact protection. We don't see them, but side impact beams in the doors are a mighty reassuring presence.

The interior didn't disappoint either: well upholstered and, unlike its

145

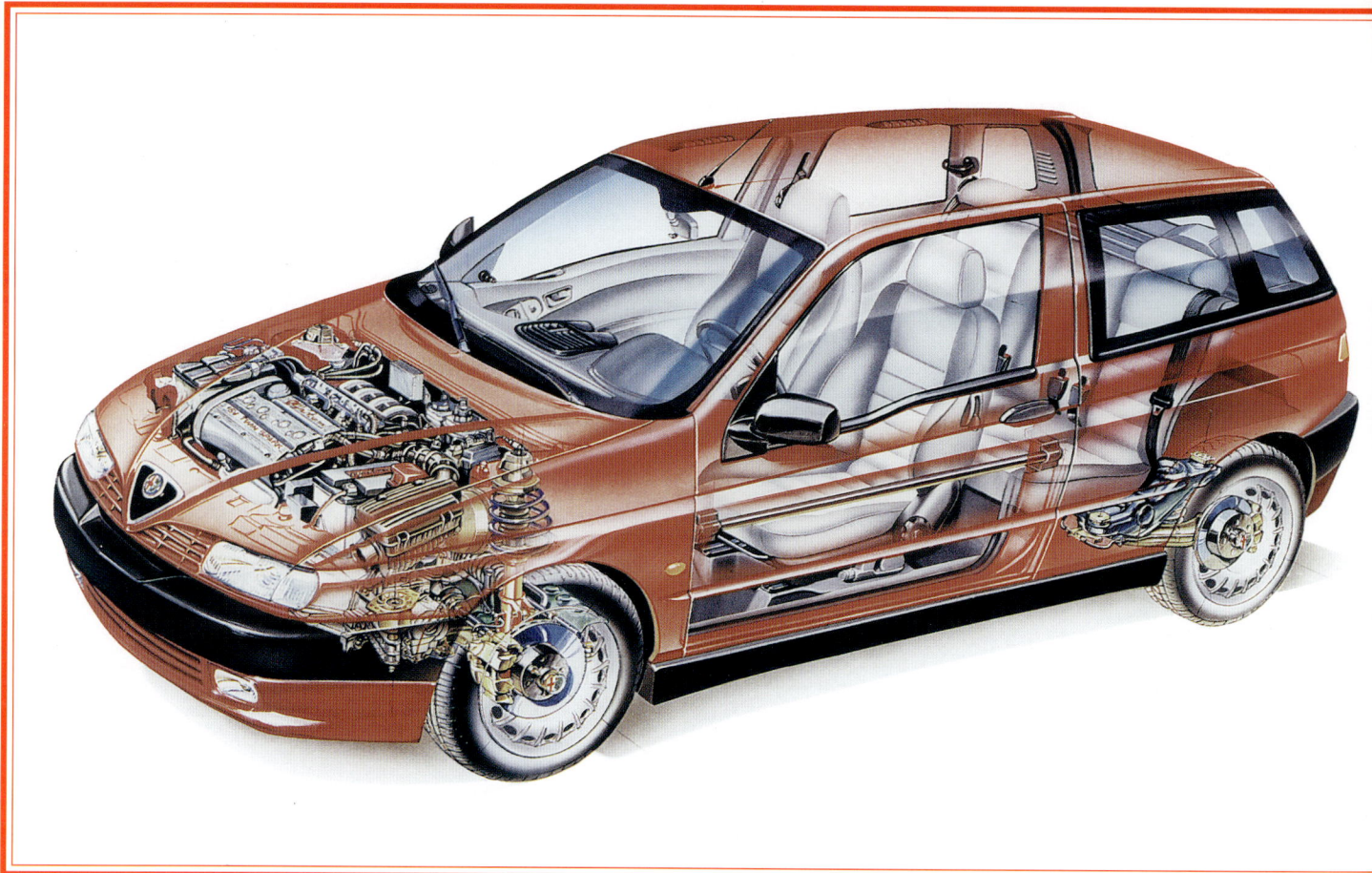

The compactness of the 145 1.8 T. Spark powertrain and front suspension system is evident in this ghosted illustration, as is the side impact beam in the door. Also identified are the areas of plastic trim.

predecessors, the Recaro-style seats seemed supportive as if meant for the job. The front seats had fold-down backrests, with a sliding mechanism on the front passenger seat to facilitate entry into the rear of the car, and the back seat folded away to create additional luggage-carrying capacity. Instrumentation was better co-ordinated and the controls a pleasure to use. The instrument binnacle ahead of the driver contained the speedo, rev-counter, water temperature and fuel gauges,

The seats in the 145 Cloverleaf have two-tone upholstery, with a coarser weave in the central area, and cabin ergonomics are generally very good. Across the car you can see the distinctive dip in the line of the door capping.

Under-bonnet shot of the engine bay of the 146, revealing its transverse-mounted, 16-valve, 1.8 Twin Spark motor. The blips on the top of the cam cover show the location of the sparkplugs, and the heatshield at the 'front' of the engine shows where the exhaust manifold lives. The battery is on the left-hand side, with the air filter and intake tucked away behind it.

and the centre console was angled towards the driver to present the radio and auxiliary instruments. The Momo wheel contained an airbag. The most obvious styling innovation was the cutaway of the panelling ahead of the front seat passenger which gave a much greater sense of spaciousness, particularly in the front.

By 1995, the flat-four's days were numbered. Strangely enough, I've never been that much of a fan of this engine, which is odd for someone who likes 911s and has run Beetles and Citroën GSs in the dim and distant past. I put it down to a blind belief in the aluminium twin-cam as the rightful occupant of an Alfa Romeo engine bay.

Anyway, the first 145 model to receive the straight four Twin Spark was the top-of-the-range 2.0 16V Quadrifoglio introduced in 1995. As we've already seen, these state-of-the-art engines were developed by Fiat and utilised a cast-iron block that was stronger than the old all-aluminium unit. The Twin Spark engine had four valves-per-cylinder, variable valve timing, hydraulic tappets, the latest generation Bosch Motronic engine management system, a steel crankshaft with eight counterweights and contra-rotating balancer shafts. The maintenance factor was a critical aspect of running the old aluminium twin-cam, but the new Twin Spark engine required no such attention, other than oil changes between 100,000km (65,000 mile) service intervals. The modular family of Alfa Twin Spark engines was built at the company's Pratola Serra plant at Avellino, south

of Naples, which came on stream in 1997. Engine capacities ranged from 1.4- through 1.6- and 1.8- to 2.0-litres. The 145 2.0 16V Quadrifoglio developed 150bhp at 6200rpm, and was good for 210kph (130mph). It contrasted with the rest of the range in having a fuller air-dam and body mouldings front and rear, plus side skirts and six-hole telephone dial wheels.

The 33's successor

One result of the changing fortunes of the company was its new-found self-confidence, manifest in the production of a five-door variant on the 145 theme. Launched in May 1995, the 146 could be considered a 145 for the family man, with straightforward access to the rear seats and revised luggage space. This, to me, was more like the successor to the Alfa 33. It wasn't as aesthetically brilliant as the 145, nor as hunky as the larger 155, yet there was nothing not to like about it. If that sounds like I'm damning it with faint praise, then I guess it's because we were spoilt for choice. There were already plenty of Alfas to choose from, and this one wasn't especially

outstanding. There was still a wedge-like stance, but also something of the jelly-mould, which made it seem a bit tubby. Like the 145, there was a prominent swage line that started at the front and rose along the flanks to end at the high, truncated rear end that qualified neatly as a Kamm tail. The rear valance was also markedly concave, and the 146 had the benefit of a rear window wash-wipe system.

Like its sister car, the 146 had the passive safety features that included side impact beams within the doors, and much the same mechanical specification, including ABS brakes with ventilated discs at the front, independent suspension by means of MacPherson struts at the front, cast iron wishbones, mounted with differential stiffness bushes, negative offset kingpins and stabiliser bars, with coil springs and dampers all round. There were longitudinal trailing arms mounted to the body via a cushioned subframe at the rear, plus an anti-roll bar. The identical range of engines powered it, and the interior was essentially the same as for the 145, but without the need to access the rear compartment in the same way.

Being a four-door car, the 146 can be seen as the true successor to the Alfa 33, which was always a four-door model - or estate. The 146 shown here demonstrates the original design concept of the 'beak' bonnet and simple apertures either side of the Alfa shield. It also has rectangular driving lights in the valance rather than the later circular ones.

By 1998, the 146 was available in three forms, with the 1.6 Junior as the entry level model, the 1.8 T. Spark and the range-topping 2.0-litre ti. Call me cynical, but when the special editions start appearing, it's a fair bet that the standard model's not selling all that well. The Junior was offered with substantial discounts, and much was made of its racey body kit and all that jazz. It was marketed as a descendant of the Giulia 1600 GT Junior coupé of 1974, and featured its own logo, distinctive 15in alloy wheels and colour coded side-skirts, door handles and rear wing. Pretty good value, in fact, and a sensible choice for the short-haul family. The 1.8 T. Spark (note the abbreviation) was the Plain Jane of the 146 range, although it was perfectly well appointed, while the 2.0 ti model had more or less everything that the other two cars did, but with an obvious improvement in performance. It also came with different multi-hole alloy wheels,

That rear-end with its shallow concave shape is reminiscent of the Kamm-tail models from back in the 1960s - the Coda Tronca Giulietta SZ and Giulia Super - and the top-spec 146 is also identified as the ti for turismo internazionale (in lower case initials, note). With four doors and a little more capacity than its 145 sibling, it probably is a better bet for family holidays.

148

The 146 was more than just a variation on the 145 theme, being a full four-door saloon with a rear hatch accessing the luggage boot. Originally available with the flat four boxer engines, this is the post-1999 facelifted car, showing the more complicated side grilles and five-hoop alloy wheels.

sill (rocker panel) covers and a boot spoiler.

Facelift

In mid-1999 Alfa Romeo revamped both 145 and 146 model ranges and they were offered as 1.6 TS, Cloverleaf and TI versions. The Junior versions were replaced by the 145 and 146 1.6 TS 16v, which, in both cases, retained the Junior's sporty looks but not the badge. The model line was completed by the 1.8 TS and 2.0 TS in Cloverleaf and TI formats.

The now-familiar Twin Spark 16-valve engines came in 1.6-, 1.8- and 2.0-litre capacities, and featured variable valve timing, variable geometry intake systems in the 1.8 and 2.0 versions, together with four valves-per-cylinder and Twin Spark ignition. Gains were improved flexibility and mid-range torque derived from effective fuel management and better power delivery. Piston cooling was assisted by oil jets, and the ignition was a phased sequential type with individual integrated coils for each cylinder. The output from each coil was directed to two different cylinders. Low maintenance features included hydraulic tappets, automatic timing belt tensioners and long-life sparkplugs.

Whilst not new to Alfa Romeo, the latest variable valve timing system made use of the latest electronic engine management system. The phase shift was actually less than in the past, a matter of 250 rather than 300, but it moved to provide full overlap at as little as 1800rpm when maximum torque was required, and eliminated overlap

at either full power or idling. Over 90% of peak torque was available from between 2500 and 3000rpm, depending on the version. The 1.8 litre Twin Spark 16V also had variable geometry intake manifolding. The intake manifold contained a glassfibre-enriched nylon system of moving parts, which channelled the air into 380mm or 560mm ports according to immediate requirements. This was called 'intelligent variability', and was controlled by the same electronic unit that regulates the engine's ignition and fuel injection functions. This system exploited ram-effect and acoustic resonance phenomena to ensure that the cylinders were always filled appropriately. The dual advantages were, firstly, that power and torque output at intermediate rpm were uprated, and, secondly, it optimised the smoothness of the entire drivetrain so that driveability and overall engine efficiency and fuel economy all benefited. The 2.0-litre Twin Spark embodied all these features, and also contained the Vibrodine twin balancer shaft system to minimise vibration.

Engines were fitted with stainless steel exhaust manifolds and exhaust gas recirculation equipment, three-way catalytic converters with heated lambda probes, and all units complied with EC Stage 2 exhaust emissions and noise standards. All drove through a five-speed transverse gearbox. All versions were fitted with the 'quick' high ratio steering rack, with 2.1 turns lock-to-

lock, as fitted to the late model 155, the GTV and Spider, which was hydraulically-assisted and speed sensitive. The steering column featured a telescopic lower section and collapsible steering wheel. The braking system consisted of 284mm ventilated discs at the front and 240mm discs at the rear, with vacuum servo-assistance. All 145 and 146 models were equipped with Bosch 2E ABS anti-lock brakes as standard.

Mechanical upgrades are fair enough, but bodywork facelifts are another matter. I'm not altogether sure that change for the sake of it necessarily improves the product, and this may be a good instance of that. What I'm saying is that the heavy bumpers didn't do anything for the Alfetta saloon, nor were the plastic additions particularly advantageous on the GTV. It isn't always thus: the 33 in its final incarnation looked a whole lot better than when it first came out. Anyhow, the changes made to the 145 and 146 included new radiator grilles, new shape body-coloured bumpers, with black inserts for the 1.8 versions while the 1.6, TI and Cloverleaf had body-coloured inserts. There were new round fog-lamps in the front valance instead of rectangular ones, and a revised colour range for all models.

In the cabin were to be found chrome door handles, instrument bezels and handbrake button. The 1.6 TS gained a sunroof and single CD tuner in lieu of a radio/cassette player as

standard equipment, while the 1.8 TS had air conditioning and single CD tuner as standard, and headlamp washers and leather upholstery as optional extras. The front seats featured an anti-submarining squab design, which prevented the driver or passenger from sliding beneath the seatbelt in a heavy impact. The 2.0 Cloverleaf and TI versions got air conditioning and new alloy wheels as standard, and the 10 disc CD autochanger was retained. More crucially, driver, passenger and side airbags became standard on all 145 and 146 versions.

A comprehensive on-board fire prevention system was installed, including an inertia switch that cut out the electric fuel pump in just a few thousandths of a second, along with three fuel-flow stop valves. A flameproof fuel tank was positioned in a protected part of the car, and flame-retardant upholstery material was used to comply with the stringent requirements of US 302 safety regulation standards. The manufacturer also claimed that only fully recyclable materials were used in production.

As with all contemporary Alfas, the prime security feature was the Alfa CODE ignition key, which activated the on-board immobiliser system through a transponder concealed in the key grip, sending a signal to an aerial coiled around the ignition switch. When the transponder emitted a unique code, it was picked up by the aerial and activated the immobiliser control unit, freezing the engine management computer and preventing the engine from being started without the correct key.

In addition, all models, including the 145 and 146, were fitted with a sophisticated remote controlled alarm system as a deterrent against forced entry and, with an Ultrasonic facility, protection from glass being broken. Typically, all windows were etched with the vehicle's identification number and eight digit vehicle chassis number prior to sale, and each vehicle was automatically registered with the NVSR. The alloy wheels were also fitted with anti-theft bolts.

Like all modern Alfa Romeos, the 145 and 146 models came with a three-year, 160,000 mile mechanical warranty, in addition to an eight-year corrosion protection warranty and a three-year paint warranty. Astonishingly, there was no price increase (in the UK at least) for the 145, and the facelifted 146 price was even reduced.

Meanwhile, waiting in the wings was a replacement for both cars, rationalised as a single model - the 147. It would be a three- or five-door hatch, featuring similar rear-end styling to the 145, with aggressive frontal treatment, and scheduled for launch in 2001. The entire range of Twin Sparks would be available, plus a Selespeed option. The design was signed off by Autumn 1999, and it was unmistakably an Alfa Romeo.

13
THE THINKING MAN'S BERLINETTA

The 156 is such a fabulous looking car that it's almost a shame it wasn't launched in 2000 in order to signify a clean break from the last century. No matter; there are few enough cars that have such a timeless quality about their design.

The 156 has such a commanding presence on the road, evoking images of sports saloons from years past - the round-edged Giulietta Berlina certainly, possibly the Giulietta Sprint and the bulkier 1900, and even the rotund little Giulietta SZ coupé. I think you can also conclude that the *Freccia d'Oro* and *Villa d'Este* from the late 1940s were present in Walter da Silva's reference book when he was looking for inspiration for the 156. And that's just covering the stance and dynamics of the overall shape. It's also telling us that Alfa's halcyon days were forty or fifty years ago, when the Alfetta 158 and 159 were the Grand Prix cars to beat, that car styling then was lavish and luxuriant, and that we can have some of that now. More fundamentally though, the 156's lack of rear door handles is a conscious effort to create the illusion it's really a *berlinetta* (coupé), which, by implication, is somehow faster and therefore more exotic than a mere saloon.

Get into the detailing and you

In 1997 came the 156, the model that would win Alfa probably the widest acclaim of any car it had ever produced, except possibly the long-running 105-series Spider.

151

Arese 9'82

come up with many more intentional references to classic Alfas. The polished aluminium front door handles are deliberately made more obvious to distract your attention from the rear access points in the C-pillars, and at the same time they refer directly to that icon of Alfa saloons, the Giulia Super. The panel fit is so good that the rear door shut lines are barely discernible. At either end of the car, the light clusters have a compressed, almost minimalist, aspect that also comes across as high-tech - small yet powerful - which was a feature common to most cars that started life at Centro Stile in the 1990s. In order to further the impression of a sporting car, the 156's bumpers were reduced to thin rubbing strips at the apex of the bulbous valances; because, after all, racing saloons have pretty much always had their bumpers removed to save weight.

Number plates are an eyesore, there's no getting away from it. But there are ways to deal with them, particularly at the front of the car. Cars used in competition used to have adhesive plates stuck to the front of the bonnet or the

most vertical available surface. Another solution practised by Alfa Spider drivers is to position the number plate out of the way of the radiator, ducting to one or other side of the car. Some people use a motorcycle-size plate, which, as I've found, can cause you to be pulled over and ticked off by a policeman. There's a cosmetic reason for this as well as a practical one. The original Lotus Elite was sold with its registration letters screwed onto the grille mesh within the aperture, so as not to spoil aesthetic lines, nor compromise cooling or precious aerodynamics. The designers of the 156 picked up on this issue and made a feature of it, keeping the wretched plate out of

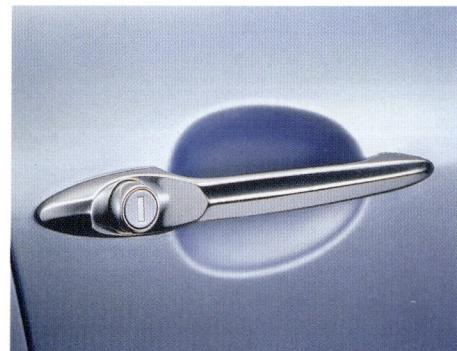

Attention to details such as the retrospective polished aluminium front door handles made the 156 a treat to use and to look at.

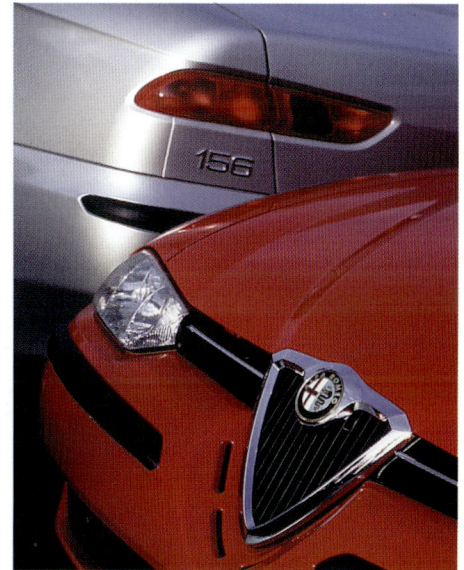

Slender rear light clusters integrate well with overall styling, and come in two abutting sections; one on the wing and one on the bootlid.

With the 156, the Alfa shield was stretched a bit, recovering the proportions of former years. The slits around the lower part of the shield are evocative of models from the 1930s, like the 6C 2500.

the way of the Alfa shield by positioning it over to the left of the car.

And that Alfa grille. It went all squashed for a while - see the 145 - but for the 156 it's been elongated to bring it more in line with traditional proportions. And just to remind you of another idolised Alfa, this time the Monza, those slots that surround the lower section of the shield in the front valance lead you straight back to the 1930s. As a further subtle twist, the top of the Alfa shield is slightly elevated to merge with a little ridge that rises from the bonnet.

Tech spec
The 156 was launched in 1997 to universally rapturous acclaim, and was promptly awarded the coveted Car of the Year accolade. In its accustomed way, Alfa Romeo offered engine options to suit most people's requirements, with the 1.8- and 2.0-litre Twin Spark models and the 2.5-litre V6. By 1999 the five-cylinder 2.4 JTD turbodiesel engine that uses the fuel-efficient Unijet common-rail direct injection system was also on offer.

The smaller capacity units were essentially the same cast iron-block 16-valve twin-cams that had seen service in the 155, and again, only the 2.0-litre version had the balancer shafts. It

mattered not a lot, though, as the 1.8 was still streets ahead of the old 1750 unit of yesteryear in terms of performance, if not character, perhaps. One of the visual aspects of the new Twin Cam engine was the enclosure of the sparkplugs, which was something that even the least mechanically-minded owner could relate to on the old engine.

This is the transverse-mounted, four-cylinder, 2.0-litre Twin Spark engine, illustrated here in the 156 Selespeed model. It was a sweet-sounding power unit, revving eagerly well up into the higher reaches of the rev band.

Now, each combustion chamber in the Twin Spark motor has two platinum spark plugs, of different sizes and performing different roles. You take it for granted that they're there and doing their job, which is another sign of the times: longer intervals between servicing, less DIY maintenance, and, most importantly, reliable mechanicals. As stated earlier, the advantage of a twin spark system is that there's more efficient combustion. The larger sparkplug is located centrally, and fires the compressed charge at the beginning of the power phase, while the smaller one is offset at one end of the chamber and sparks 360 degrees later at the end of the exhaust phase. The dual benefit is that emissions are reduced, and the catalytic converter is protected because there's no unburned fuel reaching it. Each cylinder has its own direct-ignition coil, and the sparkplug leads are so arranged that the output from each coil is directed to two different cylinders.

Because four-cylinder engines are inherently less well-balanced than sixes, Alfa Romeo builds two contra-rotating balancer shafts into the 2.0-litre Twin Spark engine to iron out the imbalances caused by alternating moving masses. The crankshaft carries eight counterweights and a torsional vibration damper, whilst the cast-aluminium alloy sump adds stiffness to the engine and gearbox assembly.

Unlike the 155, the top-of-the-range 2.5-litre V6 engine was endowed with four-valves-per-cylinder for the 156, and that lifted power output to 190bhp, which was close to that of the old 12-valve 3.0-litre unit. Having driven a V6-engined 156, courtesy of Lindfield Norwich, I wouldn't care to draw any other comparisons, as the two models are very different. If you want a contrast here, though, I'd cite the benefits of the slick six-speed gearbox of the 156 against the relatively ponderous transaxle of the '80s car.

The build quality of the 156 was exemplary, with rear door shut lines barely discernible, giving the impression - from a distance, at least - that the body was that of a two-door Berlinetta.

The six-speed 'box almost begs you to play around with shifts, which you do, just for the novelty of it, and it's a return to the knife-through-butter values that used to be ascribed to the Giulia Super. The 156 V6 2.5 has a sophisticated engine management system, too. Impulses from its electronic fly-by-wire throttle are transmitted directly by a control unit built into the engine, which regulates the rate of fuel intake by constantly monitoring the engine's rpm, the car's speed and whatever gear ratio is in use.

But does the reality measure up to the expectations fostered by the image? It's front-drive after all, so how can it provide the same kind of driving experiences you'd get in the classic forebears? The answer is that it doesn't disappoint, and while you can't compare the two motive experiences, modern with classic, the 156 is exceedingly competent as a car of its time. I came to this somewhat regrettable conclusion after driving a 156 Selespeed one day and, by coincidence, a Giulia Super the next day, while researching this book. They are like chalk and cheese. The former is planted firmly on the road, turn-in and handling as absolutely sure-footed as

Sherpa Tensing in the Himalayan foothills, the second is a frisky pony up for a romp. Perhaps the answer is that you could live with the 156 on an everyday basis, while the Super provides weekend stress relief. What am I saying? This is heresy! Did I not run a Super as my regular transport for five or six years? Ah yes, but that was a whole ten years ago, and the 155 has, I'm afraid, converted me to the easy ways of a fast front-driver.

All fingers and thumbs
The Selespeed shift system was introduced in the UK in April 1999, amid a certain amount of hype that

The controls of the 156 have much more of a 'hewn from solid' quality about them than previous models. This is the Selespeed version, introduced in 1999, with the gearshift buttons on the steering wheel boss: up-shifts to the right and downshifts to the left.

here, for the benefit of race fans, was Formula 1-style gearchanging. It wasn't quite like that in practice. But whatever, this is how it worked. 156s so equipped had two buttons each side of the steering wheel boss, as well as a sequential gearstick on the transmission tunnel. There was also a fully automatic transmission function called City, which came in at the touch of a button. I was lent one for the day by my local dealership, and salesman Alastair Kerridge took his time to explain how to operate it, much as if I was a regular punter. It's as well he did, as it's not something you can just pile into and whizz off up the road with.

In principal, once the car is rolling, the manually actuated five-speed transmission allows for gearchanges to be made without the driver's hands ever having to leave the steering wheel. There's no clutch pedal, and really no need for the driver's foot to leave the accelerator. The computer-operated Selespeed system releases the clutch automatically and brings the engine and gears to the same speed, then re-engages the clutch, all in a fraction of a second. You then have the choice of changing gear either via the push-button controls on the steering wheel, or the sequential gearstick. The push button system changes up with each thumb press on the right one, identified by a '+' sign, and downshifts with each press

155

on the left one, which is identified with a '-' sign. The gearstick works in the same way as the buttons. As with a sequential shift, you simply nudge it forwards to change up, and backwards to change down.

The system was developed by Alfa Romeo in conjunction with Magneti Marelli, which manufactures the transmission package. It has a fully electronic 'drive-by-wire' throttle, which means that the software can summon up exactly the correct engine rpm to ensure smooth shifts during any downchange as the electronics match engine and road speed. That's why the clutch pedal is redundant, and, theoretically, it should make for swift, crisp gear changes provided you feather the throttle momentarily during shifts.

User chooser

To put it a more technical way, Selespeed is a manually operated, electro-hydraulic gearbox with an automatically activated clutch mechanism. However, it's up to you as driver to elect which gear you want to be in. But, if you hesitate too long, the gearchange is made automatically. The advantage of Selespeed is that you really can shift gears neatly and accurately while keeping both hands on the steering wheel, without having to operate a clutch, and this could be a real boon on a fast back road with

lots of corners. It can work in automatic as well as semi-automatic mode, which I used when doddering around in a town that I didn't know. You simply press the button marked 'City' on the transmission tunnel to activate automatic mode. The steering wheel or gearstick controls aren't required when this mode is selected. It meant I could leave the car entirely to its own devices while looking out for street names, juggling a map, and so on. Three actuators intervene to carry out the actions that are normally performed during conventional gearshifts. The first controls the clutch, the second controls engagement, while a third controls gear selection. A fourth actuator is linked to the engine's electronic throttle and manages torque flexibly, according to input from the gearbox control system.

The Selespeed system is based on an hydraulic power system, incorporating an electric pump that takes oil from a reservoir to an actuator, generating pressure in the circuit and providing the energy required to move the clutch engagement and disengagement levers, to select gears, and operate the clutch itself. Because the system needs energy from start-up, the pump begins to operate as soon as the driver's door is opened, and the circuit will have already reached the required pressure by the time you're ready to drive away. Fuel supply is governed by a dedicated electronic control unit, manufactured by Magneti Marelli, while the control unit communicates continuously with the Bosch Motronic unit that manages the engine. It's responsible for process-

ing the driver's input and converting it to gearchange commands, using information such as accelerator pedal position, vehicle speed, engine rpm and torque input. The system is also supplied with data by sensors and potentiometers located at specific points of various driveline components, which it also uses to monitor the status and condition of components. A VDU inset into the rev-counter indicates which gear is engaged, or whether it's in City mode, or any faults in the system.

Starting out

To get going, you press the brake pedal and select the appropriate gear: first, second or reverse. You use the gearstick on the transmission tunnel for this, because the steering wheel controls aren't active at speeds below 6mph (10kph). You then release the brake and accelerate, and from then on it's possible to change gear using the steering wheel controls or the sequential stick. (The latter takes priority if both controls are operated at the same time.) You can keep the accelerator pedal depressed during gear changes, although it's smoother if you lift off. In the event of an emergency stop, the system detects the rate at which engine speed is dropping and automatically changes down through the gears. On the other hand, it's necessary to keep half an eye on the VDU when shifting down to keep count of the number of times the left-hand button has been pressed, sending the revs sky high as you inadvertently find yourself in first or second.

If you prefer to be less proactive about your progress, as you approach a crossroads or traffic lights, for example, Selespeed will slow the car, automatically shift down and release the clutch to prevent the engine stalling. If you move off again without stopping completely, the system automatically engages the best gear for pulling away. Otherwise, it automatically engages first gear when the car stops.

There are one or two other idiosyncrasies to bear in mind, which is what Alastair Kerridge was keen for me to be aware of. To engage neutral with the car at a standstill, you press the brake and push the gearstick to the right. If the accelerator and brake pedals remain inactive for one minute, the gearbox automatically selects neutral. It won't let you select neutral travelling over 25mph (40kph), and a gear must be engaged before the engine can be switched off and ignition key removed. If the key is turned off with the gearbox in neutral, a warning buzzer sounds. The system de-activates itself two seconds after receiving a zero speed signal from the engine and the gearbox input and output, having saved operational and test data in its control unit's permanent memory. The system automatically moves into neutral when the car is restarted.

It also tells you when a gear change is impossible; for instance, reverse engagement is indicated by an acoustic and visual signal, and another signal notifies the driver when the engine is switched off and the gearbox is left in neutral. A warning light also comes on when transmission oil pressure is too low, and an intermittent sound warns when the clutch is slipping. It can also test automatically for faults in its own componentry. You can tell a Selespeed car by its boot badge and the set of exclusive 16-inch alloy wheels which are fitted as standard.

Q-cars

An automatic transmission version of the 156 2.5 V6 24V, mysteriously known as the Q-System, went on sale in the UK in April 1999. Q-System is manufactured by the Japanese company Aisin, and is a 5040 LM unit with four forward gears and reverse. Its torque converter is equipped with a lock-up function, that's to say, a clutch mechanism that bypasses the converter when third gear is engaged, in order to minimise power losses, reduce fuel consumption and lower transmission oil temperature.

Like the Selespeed, it's effectively two transmissions in one. A central console-mounted gearlever gives you the option of fully automatic transmission when positioned to the right, or a four-speed clutchless manual option when moved to the left. The thinking behind the Q-System is that all the hard graft is taken out of city driving, while giving the driver something to play with out on the open road.

Further refinements include three different pre-programmed driving modes, identified as City, Sport, and Ice, available when the lever is positioned to the right in Automatic. You select one of these by pressing two buttons at the base of the gear lever, and the selection comes up on a VDU inset into the rev-counter. If clutchless manual control is preferred you just move the lever to the left, where it slots into the traditional four-speed H-gate arrangement. With the lever to the right, you select Automatic, with the four regular options of Park, Reverse, Neutral and Drive available in the normal way by moving the lever forwards and backwards. From the 'D' (Drive) position, you can shift the lever over to the left and access the four-speed H-gate manual option as you like. However, reverse is accompanied by an acoustic warning signal and can only be accessed in Automatic.

Characteristics of the system are predictable. In the Automatic settings you get low rev gearchanges in City mode, swifter acceleration, engine braking and gearchanges at much higher engine speeds in Sport mode, and the Ice ratios reduce the chance of wheel spin by using second gear to start off in, promoting a smooth getaway. Manual mode returns full control to the driver, but the system nevertheless prevents you selecting too low a gear at too high an engine speed.

Like Selespeed, the brake pedal must be depressed in order to engage any of the gears when the car is stationary. A locking shift-ring immediately below the gearknob has to be raised while your foot remains on the brake pedal so that the gearshift can be moved out of 'Park' and into 'Drive' or 'Reverse'. With the gearshift in 'D', the system automatically engages the most suitable gear according to the driving mode selected. City and Sport modes are selected by pressing the

The appropriateness of a diesel engine in a sports saloon is open to question, and ownership thereof perhaps a matter for one's conscience. But Alfa Romeo claims that the 2.4-litre JTD in its 156 can push it along to 126mph (200kph) and from 0-to-60mph (100kph) in less than 10 seconds. The single benefit has to be that you get 41mpg. But if you can afford a 156, why are you worried about saving 10mpg when you could be running the zinging Twin Spark or spine-tingling V6?

two-position button at the base of the gearshift, in either C or S. Ice is accessed by the second button and cancels out the other two modes. It's only active at speeds of less than 28mph (45kph) and automatically cuts out at 50mph (80kph).

Another fail-safe aspect of the system comes into play when the car's stationary. You press the footbrake and the ignition can be turned off and the gearshift moved into 'Park'. The handbrake is applied and the solenoid will be released after a couple of seconds' delay, so you can then remove the ignition key. Leave it in any other setting and the key can't be removed. An acoustic warning sounds as the driver's door is opened, reminding you that it's in gear and the key is still in the ignition. Clever stuff, but it's a sign of the times. Do we really want this sort of mollycoddling?

Oil be damned

Introduction of the 156 coincided with Alfa's innovative diesel engine, which was the first in the world to incorporate a Unijet system and was available in the 156 2.4 JTD model. On account of the fact that 30 percent of Alfa Romeo's European sales involve diesel powered cars, it's as well to expand on the specification. As I've made fairly clear, I've got somewhat blinkered views on the place of the diesel engine in sports saloons, but here goes.

Alfa claims that given its 136bhp and 224lb.ft of torque, the Alfa 156 2.4

JTD should be regarded as a performance diesel, as it puts many petrol-engined cars to shame. Helped along by a turbocharger, it can do 126mph (200kph) and 0-62mph (100kph) acceleration in 9.5 seconds, which is, well, OK I suppose. We used to think that anything that could do over the ton was good thirty years ago, and 0-to-60mph (100kph) in under 10 seconds wasn't to be sniffed at. The Alfa 156 2.4 JTD is powered by a 2.4-litre,

five-cylinder turbodiesel engine, featuring the high-pressure direct-injection system known as Unijet. It's a 'common rail' system, designed and initially developed within the Fiat Group, with input from Magneti Marelli and productionised by Bosch.

So what's new? Well, compared with conventional injection systems, Unijet offers better performance and a quieter engine. As Jennifer Aniston would say, here comes the science. In the Unijet system, injection pressure is independent of engine speed and

Launched in 1999, the 156 2.4 JTD was powered by the company's innovative new diesel engine, a 2.4-litre, five-cylinder turbocharged unit. What made it special was the high-pressure, direct-injection Unijet 'common rail' system, designed and developed by Fiat and Magneti Marelli, and produced by Bosch.

accelerator pedal position, since the injection pump generates pressure on a cumulative basis. And since both the pump and injectors are electronically controlled, injection pressure and fuel quantities injected can be optimised at every point on the power curve. This makes it possible to combine very high injection pressures with the electronically controlled delivery of minute quantities of diesel fuel in order to achieve pre-injection. This is known as pilot injection, which reduces combustion noise.

In addition, high pressure injection significantly reduces both fuel consumption and exhaust fumes, while pilot injection creates the ideal temperature and pressure conditions inside the combustion chamber for the actual combustion to take place. Preheating the combustion chamber drastically reduces the pressure gradient, which is the source of uneven combustion; hence the racket that's typical of conventional direct injection engines. I must say that the two-stroke supercharged six-cylinder Foden diesel in our boat roars like a vintage race engine on full throttle. But that's another story ...

Common rail system

The Unijet system on the Alfa Romeo 2.4 JTD consists of a small pump immersed in the fuel tank, which delivers diesel fuel to the primary or high pressure pump driven by the timing belt. This continuously pushes the diesel fuel into an accumulator tank, or 'rail', which always contains pressurised oil, whatever engine speed

and load - whatever is demanded by the accelerator pedal. A pressure sensor on the rail and a pressure regulator on the pump respectively monitor oil pressure inside the accumulator, and adjust it according to instructions from the electronic control unit. The pressure sensor reads the bar level inside the accumulator, and if that level is too high, it passes the information on electronically to the pressure regulator - an hydraulic valve - which returns excess fuel to the tank. Fuel pressure is thus continuously variable, and ideal pressure can be selected for any individual point on the engine's power curve. The higher the pressure of the fuel entering the injector, the finer the fuel spray produced, which makes for a better mix of air and fuel and a more complete combustion.

To reduce noise levels, Unijet's pilot injection uses an electro-hydraulic injector system governed by an electronic control unit, which raises combustion chamber temperature and pressure, just as each piston reaches TDC, and prepares the chamber for combustion proper. Through pilot injection, the heat release curve at the moment of main injection is very much less steep, and the temperature and pressure peak is also at a lower level. The same amount of energy is produced, but it's this progressive delivery that drastically reduces noise. So what this tells us is that the common rail system enhances combustion efficiency and provides better performance, while pre-injection promotes quieter combustion, easier cold starts and reduced exhaust emissions.

In order to optimise combustion inside each cylinder, the multi-jet injectors have the smallest possible apertures to atomise the fuel, and utilise a spiral shape for the intake port inside the cylinder head to optimise air swirl. A combination of atomised fuel spray and swirling air produces an air-fuel mixture that burns ultra-efficiently. Compared with a pre-combustion chamber engine, a Unijet unit of the same size as a conventional unit delivers an average 12 percent improvement in performance and a 15 percent reduction in fuel consumption: the official fuel consumption figure is 42.1 mpg.

Diesel fitter

The 2.4 JTD engine is made up of a cast iron engine block, with built-in cylinder liners, a multi-layer head gasket and a light-alloy cylinder head with omega-type combustion chambers. A single overhead camshaft drives two parallel vertical valves per cylinder, and there are no pre-combustion chambers since the entire combustion process takes place inside the chamber, which is in the piston top. Auxiliary functions are driven by a double-acting pulley. The engine uses a variable geometry Garrett VNT 25 turbocharger, capable of varying its blade angle to speed up or slow down gasflow and hence the turbine, which is linked to an intercooler. A two-mass engine flywheel, with half built into the crankshaft and the other into the mainshaft, reduces vibration and makes the whole transmission system quieter and more refined. There's also

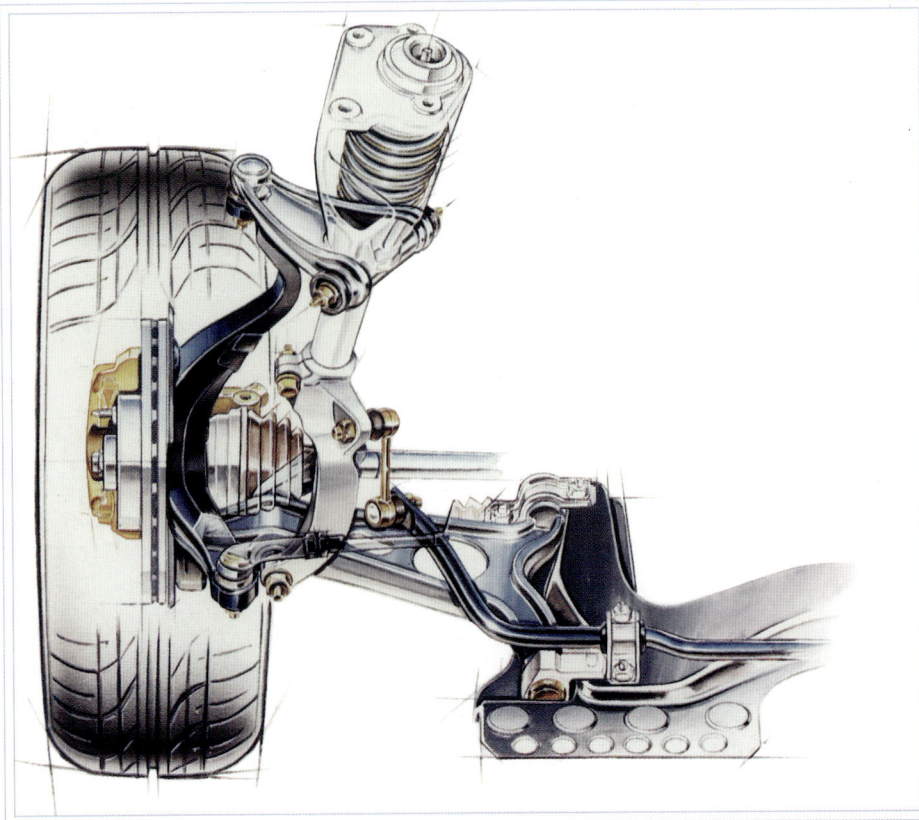

Diagram of the 156's independent front suspension, showing double wishbones, coil spring and telescopic strut damper unit, rose-jointed anti-roll bar, with driveshaft and ventilated outboard disc brake assembly with ABS.

Illustration of the 156's rear suspension set-up, independent by differential-length tie rods connected to an aluminium beam, with longitudinal reaction rods and coil spring over damper units, and an anti-roll bar connected to the dampers. Dual circuit disc brakes are outboard.

a contra-rotating balancer shaft that further reduces noise and vibration.

The front suspension of the 156 consists of the racey double wishbones, coil springs and dampers set-up, with rose-jointed anti-roll bar connected to telescopic struts. This layout is calculated to ensure an anti-dive effect, which means that the front end doesn't dip under braking or lift up when the car's accelerating. The car's rear suspension marks a bit of a departure from traditional practice, comprising MacPherson struts with differential length tie rods connected to an aluminium beam, with longitudinal reaction rods, offset coil springs and dampers. The set-up is completed by a rose-jointed anti-roll bar, and the overall aim is to control those undesirable - but inevitable - motions of pitch and yaw, at the same time creating a self-steering effect that will enhance the car's stability when driven hard. The 2.5 V6 model I drove was adequately agile when pressed, despite the belief that the heavier V6 is less sprightly

We've seen that Alfa Romeo has made sporting coupés and Berlinettas since the early days, and the GTV 2.0 TS and 3.0 V6 were the offerings from 1996. The product of the Centro Stile drawing office, the GTV displayed many of the stylistic hallmarks of its brethren, like the resolution of the bonnet and Alfa shield, the slash that ran from front wheelarch to the rear of the cabin top, and, of course, the engines were the same as those in the 166.

Left - This shot of the wonderful pearlescent blue model, with its ability to change colour like a chameleon, was taken on the Alfa Romeo stand by Nathan Morgan when the 156 was launched at Earls Court in 1997. Rumours that at the time the author was asleep in the red car in the background are totally unfounded ...

The 156 may have offered the best driving position (for tall people) of any Alfa Romeo before the 166, but the wood-rim wheel and fake wood veneer on the console were not in the best possible taste. Far better was the imitation Kevlar effect available on the 1.8 model.

than the four-cylinder cars. Its turn-in was sharp, and I could clip any apex I cared to aim at with unerring accuracy. I couldn't detect any tendency to understeer, and it was a pleasure to drive quickly along fast, winding highway. The current, or let's say 1990s generation of Alfas, from 155 through Spider and GTV to 156, have this tendency to deliver a pronounced thud when you drive over serious undulations like traffic-calming humps and dips or potholes, even at very low speeds.

Everything I have to say about the 156 interior is positive. It isn't aesthetically challenging, like a TVR, and it definitely isn't austere like a Lotus Elise. Maybe it isn't sculptural in a Teutonic way. But it's a lot better than

Boot space of the 156 is much the same as in the 155, with the space-saver spare wheel hidden under the boot (trunk) floor. The trapdoor in the rear seat back allows items to be carried lengthways in the cabin.

the 155 that preceded it. When I first tried one on the Alfa stand at the Motor Show in 1997, it was an instant feeling of pleasure, like you get when you know something's just right. Not only were the seats comfortable but the driving position was excellent at last, and that comfort extended to the rear, which is so often compromised, especially for tall geezers like me. Obviously, a similar amount of thought and attention had been devoted to the interior and exterior.

The controls were equally complimentary, with the nice sculpted Momo wheel and gearshift neatly where you want them. The principal dials were housed in twin nacelles that were presumably meant to remind you of the 1750 Berlina, while the auxiliary gauges in the central console were angled towards the driver. On reflection, there was one mistake, and that was the use of the wood veneer trimming on the centre console of the two bigger-engined models. You could argue that it does compliment the wood-rim wheel, but it looks too tacky and undermines the quality of the rest of the car. The imitation titanium trim and leather rim wheel that come as standard on the 1.8 are definitely preferable. The compact proportions of the 156 might suggest that access to the cabin is going to be restricted, but this is definitely not the case, as the door apertures are plenty wide enough for easy ingress. Access to the boot is possibly less efficient when stowing larger items of luggage, although actual carrying capacity should be adequate for most journeys.

If you want a more extreme-looking 156, your dealer can help, or at least the factory can: simply order your car with one of the various Sport Pack options, or get one or those elements that most appeal to you retro-fitted. What started with the 75's Veloce body styling and the 155's Silverstone limited edition evolved into a range of cost-optional accessories for 156 owners with which to personalise their cars. These ranged from low-profile tyres with lowered suspension and *Superturismo* -style wheels to a gargantuan rear wing and side skirts and splitters. And as if that wasn't enough, you could gild the lily even more by installing Momo-designed leather seats.

Another cosmetic treat is the strange hue in the Alfa paint chart known as Nuvola Blue, which, depending on what the prevailing light source is when you see it, can appear to be silver, pastel blue, even slightly pink or gold. When seen in isolation - as at National Alfa Day this year - it looked fabulous. But was the pale blue car viewed in the morning sunshine the same silvery pink one we saw later in the day? The cost of this chameleon-like pearlescent paint scheme adds about £1k to the cost of the car, so if you liked the idea, you'd want to get

one secondhand, perhaps?

A further evolution of the 156 platform was the Sports Wagon, unveiled in March 2000 at Geneva. You can't blame them for capitalising on the excellent dynamics, and face-that-launched-a-thousand-bank-loans facade. But I've never been convinced of the need for an estate car - unless you have dogs the size of ponies - and I know there have been *Promiscuas* , *Giardinettas* and Sport Wagons in the past. But some things are too sacred to butcher.

Parallel imports

By the time that the 156 came out, the penny (lira?) was beginning to drop with those that doubted Alfa Romeo was absolutely committed to providing a package consisting of keen styling with state-of-the-art driveability. And all for rather less of your monthly salary than it takes to service the repayments on a 3-series BMW.

It may have been coincidence, but the 156 was introduced at the same time that everyone in the UK was getting excited about buying their car abroad, thus making a huge saving on the UK showroom price tag. This is naturally total anathema to official importers, but it seemed like a way for the budding entrepreneur to make

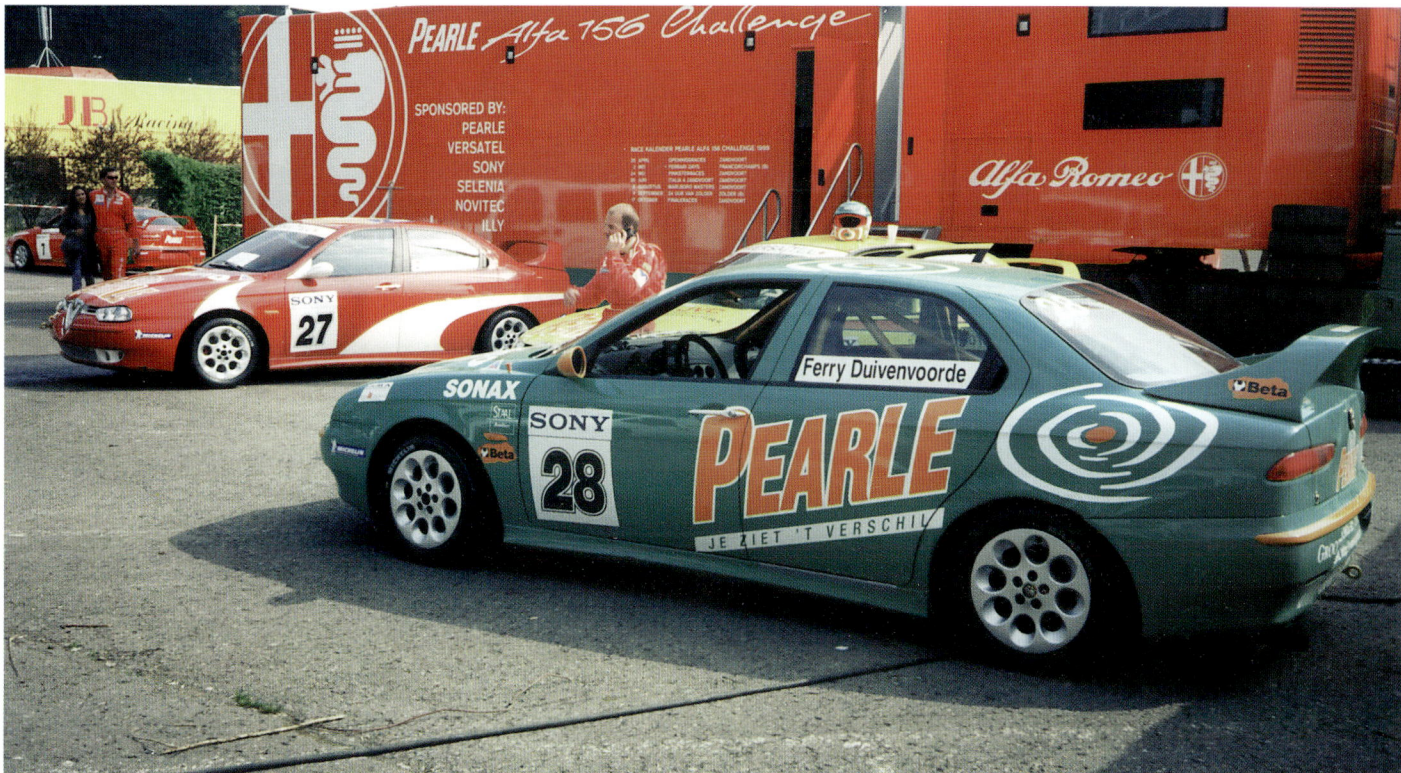

By the mid-1990s it was possible to order your Alfa Romeo ready customised with items of motor sport-influenced trim. These Group N 156 racers competing in the 156 Challenge at Spa Francorchamps are straight factory products that come complete with basic competition modifications like lowered suspension, roll cages and rear wings.

The issue of car pricing was very much a current topic when the 156 was introduced, and a number of UK buyers found they could save considerable sums of money by travelling to Holland or Italy to order their car there. Here's the Dutch car of Eric Verhage in action at Spa Francorchamps in 1999.

some relatively easy money. Foreign exchange rates with sterling were also very much on our side at the time, so I followed it up, to the extent that I located an English-speaking Alfa dealer in San Marino - where Imola is - and established that a saving of no less than £5k was possible on a V6-engined 156. The snags were that the car would be registered to me as the buyer - and therefore it couldn't be sold as a brand new car - and in order to keep the miles off it, it would have to be trailered from Italy.

Much more sensible, then, was the idea of buying in Holland, much closer to the ferry terminal. Again, I talked to several Alfa dealers in the Netherlands, and it would have been possible to undercut the UK price by similar amounts to the Italians. The registration procedure once you've bought the car is well documented - you can even access it at the *Top Gear* web site - and ensuring that you have an extended warranty rather than just the year-long pan-European one is simply a matter of buying one from the AA or RAC. So why didn't I do it, then? Time and money, really, and the apparent inability to buy the car as a dealer rather than its first owner. Plus, there are a number of specialist grey and parallel importers doing this all the time who know the system and the

procedures to follow. I have bought cars abroad in the past, but rather older ones, which haven't required the same sort of funding and daunting documentation. One day, maybe ...

Track time

Under Fiat's guiding hand, Ferrari is the marque dedicated to Grand Prix racing, Lancia the designated rally specialist and Alfa Romeo covers the Super Touring angle. There's no knowing what they'll do with Maserati, but the World GT series is one possibility. So, in the wake of the successful 155 Touring Cars of the early part of the decade, detailed earlier, Alfa Romeo went straight ahead and produced two competition versions of the 156.

The milder of the two was the virtually standard Group N car, which ran with the 2.0-litre Twin Spark engine. A great deal of the preparation of the Group N model was carried out at the factory. This included stripping out the interiors of finished production cars and installing the basic competition accessories like a roll-over cage, racing seat and six-point safety harness and fire extinguisher. Competition springs and dampers were substituted for the road-going versions, and a strut brace, an ignition cut-out and bonnet fasteners were fitted, along with

sections of aerodynamic body kit such as the rear wing and front air dam. Amazingly, all this was included in the price of the regular road car.

We, the regular punters, can get close to the experience by booking into a three-day course at the *Centro Internazionale Guida Sicura* (the so-called Safe Driving Centre) based on the Varano circuit near Parma - where the ham comes from - in northern Italy. This is run by none other than 1966 European Touring Car Champ Andrea de Adamich, whose son Gordon also raced a 155 and 145 in European events. All 50 cars in the school are front-wheel drive, and it was set up partly to demonstrate the effectiveness of this transmission layout, and also to educate candidates about how to get themselves out of trouble on a host of artificially compromised road and track surfaces. At around £1k for a couple of days, it's a country mile dearer than Brands Hatch or Silverstone, but rather more exotic.

The second of the two competition cars was the Group A 156 *Superturismo*, which, like the BTCC-winning 155 of half a decade earlier, complied with a far more liberal - if strictly policed - set of regulations. The 156 *Superturismo* power unit was based on the crankcase and cylinder head of the regular engine, but extensively modified with specially made manifolds, pistons, cams and cranks to give square internal dimensions of 86mm x 86mm. A dedicated engine management system controlled the electronic ignition and injection system, with dry-sump lubrication, and it

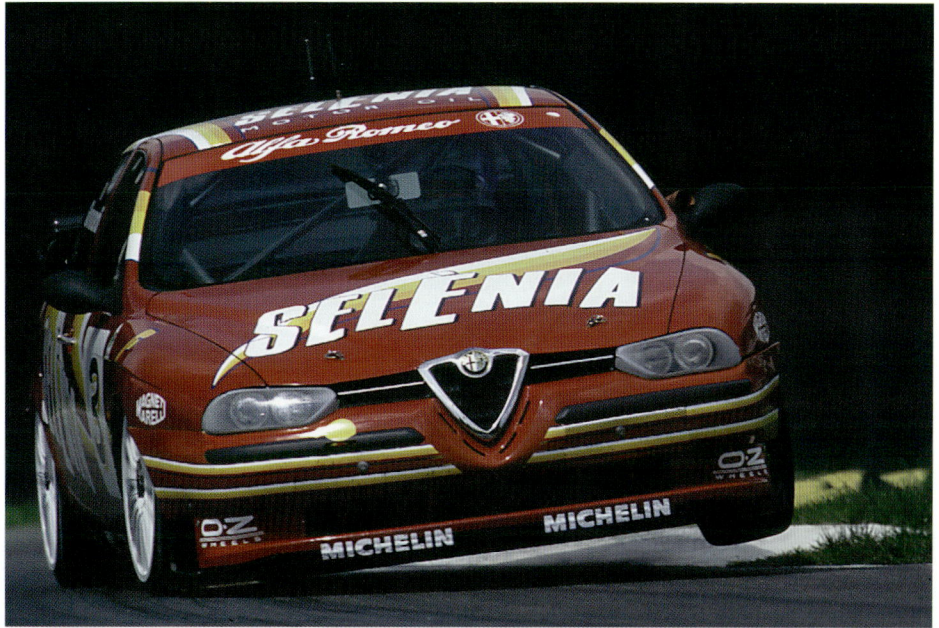

Fabrizio Giovanardi cocks an O.Z. wheel on the 156 Superturismo as he rides the kerbs on his way to victory at Imola on only the car's second time out.

developed 300bhp at 8200rpm.

Like the 155, the whole engine and gearbox unit was set lower and further back in the engine bay to improve weight distribution. Transmission was via a six-speed X-Trac sequential shift allied to a self-locking diff and viscous coupling, while the rose-jointed suspension set-up included driver-adjustable anti-roll bars. High-speed progress of the Group A 156 *Superturismo* was retarded by mighty 335mm Brembo ventilated discs with eight-piston callipers, cooled by way of yellow-coloured Kevlar ducting that framed the engine bay. The cars ran on white-painted multi-spoke 19in O.Z. magnesium alloy wheels shod with Michelin race tyres. Built-in air jacks enabled them to free-stand, as it were, in the clinical telemetry-dominated surroundings of the pits garages.

At the time of writing, the Alfa Corse 156 *Superturismo* hadn't been driven in anger outside of Italy, despite persistent rumours and predictions to the contrary. Two cars were campaigned by Nordauto under team manager Monica Sims in the Italian national series in 1998 and '99, driven by Nicola Larini and Fabrizio 'Piedone' (Big Foot) Giovanardi. Although Giovanardi won at Imola in only the car's second outing, they struggled on some circuits to hold their own against the time-served 3-series BMWs, which may explain Alfa's reluctance to release it abroad until it is certain of success.

Flagship

Having to a large extent rescued Alfa

The 166 was the replacement for the 164, and was rather more understated in design than the 156 - it didn't jump out at you in the same manner. Yet it was a handsome vehicle that maintained the arrowhead Alfa Shield bonnet styling that dips into the bumper area, its broad flanks wavering with a shallow indentation that ran the length of the car.

Romeo's reputation in the market place, the 164 was going to be a hard act to follow as the company flagship. The incoming princess was the 166, which succeeded by virtue of supreme competence rather than the kind of quantum leap in design and technology that surrounded the 164.

The 166 was a handsome car, to

be sure, slightly longer and wider than its predecessor, but lacking the extrovert personality of the 156. Its façade was pure 145, for example, and the rear was simply a different version of the elements that featured in the 156. It had a stance similar to that of its smaller sister, with a stature that harked back to the 164 - but actually

Launched at Madrid in 1998, the Alfa 166 raised the game a stage higher than the 164, placing it more directly in competition with the German cars that traditionally capped the executive segment; a natural selection rather than a choice of the heart.

it could have been any of the Type 4 cars of that generation. Again, I have to thank my local dealer Lindfield Italia for the introduction to this model, and have to say that, although it was undeniably a desirable car, it was so only on account of the quality of the interior. That, and the fact that you could get it with the 24-valve 3.0-litre V6 engine.

It was four years in the development stage, and one of the last acts of the chief chassis engineer Gianclaudio Travaglio was to hustle a pre-launch prototype around the back roads of England to see how the suspension shaped up on our particularly idiosyncratic cambers, dips and humps. Such attentiveness was a nice compliment to UK owners, but I should have thought there was plenty of bad road around Milan and Turin on which to practice.

Will it fit in your garage? A statistics check shows the 166 is 4.72m long and 1.81m wide. It was available with the 136bhp, 2.4-litre, JTD turbodiesel, 155bhp, 2.0-litre, Twin Spark 16 valve or 205bhp, 2.0-litre, Turbo V6, as well as the more appropriate 2.5- and 3.0-litre V6 units of 190bhp and 226bhp respectively.

The 166 was unveiled at Madrid in September 1998, and first seen in the UK at the Birmingham NEC in October that year. It was the final model in the company's six-car line-up, and was pitched against such newcomers as the Jaguar S-type, Volvo S80 and Rover 75, not to mention old stagers like the 5-series BMW, E-class Mercedes-Benz and Audi A6.

Cabin ergonomics of the 166 were spot-on and, as you'd expect, it was that bit more spacious than the 156. The leather-clad seats were thickly padded and wonderfully contoured,

No space-saver here. The 166 carries a proper standard-issue alloy road wheel beneath its boot floor, as well as a first-aid kit and toolkit.

166

Probably no production Alfa Romeo - and that includes the 156 - has been finished with such a sumptuous interior as the 166. The seats are fully adjustable with electric switches to ease you in every direction till you get the right position. There may be lots of black plastic still, but it comes over as good quality black plastic. And to top it all, there's a computerised satellite navigation system in the centre console.

quite as comfortable as they looked and exquisitely tailored. They were also completely adjustable electronically. The rev-counter overlaid the speedo and the fuel gauge, while the speedo in turn masked the corner of the temperature gauge, an ensemble presented as a neat visual trick within a semi-circular instrument binnacle. Other switchgear and ventilation controls were housed in the central panel, along with further instrumentation that included computerised satellite navigation aids, cruise control and a 'smart' radio that adjusted volume automatically. There was a passenger-side air bag as well as the one housed in the steering wheel. One extraordinary option, which smacked of trying too hard, was the Sport throttle, which modified

the induction system's butterfly valve opening, and which, if selected, produced a shorter throttle pedal response which was meant to make you think you were accelerating faster. This facility wasn't present on the car I tried, so I can't offer any meaningful comment.

Where the 164 was burdened by acres of dull plastic in its lower regions, the 166 had no such constraints. Bumpers were fully colour-coded, and even the rubbing strips of the 156 were absent, which gave it a purer appearance in keeping with contemporaries. The rear lights echoed the narrow strip clusters of the 156, but the sculpted indent that ran along the sides was given more prominence in the 166. Not only was this a reference of sorts to the

164, but could be viewed as a hallmark of the contemporary Alfa line-up, including the Spider and GTV. The base-model 166 2.0-litre Twin Spark got the telephone-dial alloy wheels, but different options were available for the other models in the range, the 2.5-litre V6 and 3.0-litre V6. The four-cylinder car was equipped with the regular five-speed manual 'box, while the 2.5 had the option of a four-speed ZF automatic 'box with the Sportronic shift, which was a kind of halfway house. The 3.0-litre car was available with either ZF automatic or six-speed manual 'box.

And just as the 164 shared its underpinnings with three other cars, the 166 was built on the same platform as the Lancia Kappa in a bout of inter-

167

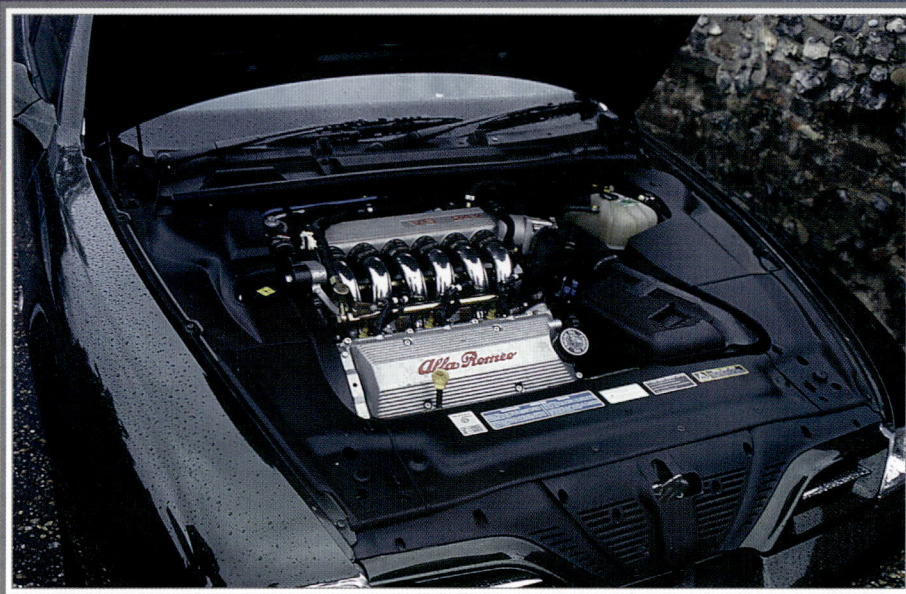

Here's how the V6 engine looks in the 166, and in this case it's the 24-valve quad-cam 3.0-litre unit. With each successive car, the engine bay takes on a more clinical appearance as the ancillaries are concealed under plastic casings.

The 166 was mercifully liberated from the acres of plastic which had marred the proportions of the 164. Instead, its sills and bumper panels were incorporated into the overall scheme in matching body colour. Only the B-post and a tiny section of glazing bar trim at the back end of the rear door were missed out.

necine synergy. The Alfa limousine's suspension set-up was very similar to that of the 156, but the bigger model's multi-link rear end was developed using a mixture of stainless steel, cast iron and aluminium components. It was also fitted with anti-slip and traction control, and ABS brakes with active sensors.

In its heyday, the 164 was selling at the rate of around 50,000 units a year, and this was the kind of capacity expected of the 166. If demand required, production could be increased to cope with a maximum of 75,000 units. The 156, on the other hand, accounted for 60 percent of all Alfa sales in 1998-99, with 200,000 units of all versions sold in its first 18 months of production.

One day I shall own one of these cars ...

14
SAVING THE SURVIVORS

There are a number of excellent Alfa restoration specialists in operation. Sadly, the man who provided the basis for a similar chapter in my 1993 book on the Giulia Sprint GT, Malcolm Morris, has since died. He was a thoroughgoing expert of the old school, whose speciality was 105-series Giulias, and a legacy of his reputation crops up in adverts for certain cars for sale, which proclaim them to have been restored by him, and which of course wouldn't be the case if they hadn't been done very well. Subsequent to doing the Giulia coupé book, Malcolm restored my GTV6 racer back to road-going status, and it looked a million dollars in ruby red *amaranto* as a result, complete with an engine rebuilt by Alan Bennett and his Benalfa team at Westbury.

As I've said, there are a number of operators whose work is excellent, like Norfolk-based Jonathan Smith who specialises in creating replica GTAs and GT Ams. Now though, it appears that Malcolm Morris's mantle has passed on to Mike Spenceley, whose Purley, Surrey, operation also specialises in 105-series Giulias. Like Malcolm, Mike has also been one of the concours judges at National Alfa Day. It also happens that Mike Spenceley is an avid fan of Giulia Supers, and I spent some time at his workshops where he ran through the major stages involved in restoring one of these cars. Evidence of his handiwork is evident in the immaculate T.I. Super of Richard Everton, and James Castle's Super that featured in the 1999 Malteasers advert (they melt in your mouth, not in

Restoration specialist Mike Spenceley, right, with James Castle, perched on the bonnet of the 2000 Berlina saloon that Mike sold to James. Anyone who's seen the Malteasers advert featuring two girls going round a roundabout in a Giulia Super may like to know that the car belongs to James, and it was also restored at MGS Coachworks.

your hand ...). Mike's own pride and joy, a Super liveried like Richard Everton's in the Jolly Club racing colours, could equally be cited as one of the finest in the country.

This Giulia Super Giardinetta *was built on 30 th January 1971 by Carrozzeria Colli in Milan, and originally supplied to the Soc. Autostrade Roma for use as a police patrol car. Its factory-fitted sunroof was positioned in order to fire on criminals during car chases. First painted 'ministry green', this rarity is pictured in the throes of restoration in Dutch blue.*

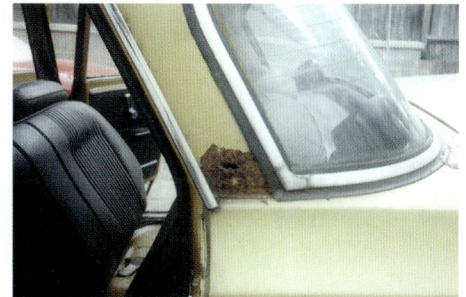

Perished screen rubbers allow water to seep behind areas like the C-pillar, with inevitable results. The structural integrity of the car is obviously compromised as a result.

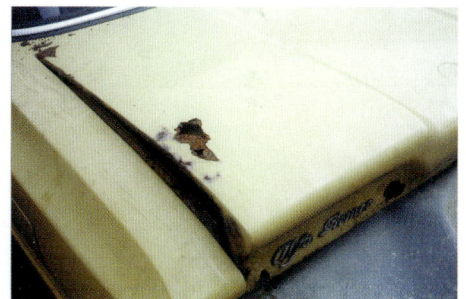

Boot and bonnet lids can attract a build-up of moisture-harbouring matter around the rims, and penetration, perforation and spalling of the metal will surely occur.

I guess we're all fairly familiar with what it takes to restore a classic car now, and first and foremost the owner has to be able to swallow hard and reach ever more deeply into the bank vaults to fund the operation. If this isn't going to be possible, you're better off, to my mind, getting a car that someone else has already paid to have restored, but for whatever reason doesn't want to keep any longer. That's by way of saying that only in exceptional cases can you hope to recoup the cost of a comprehensive rebuild - like a GTA, perhaps. The downside of this advice is that you won't have the pleasure of collecting the pristine article from the restorer, or of intimately watching its progress through the restoration process, which you can be involved in to the extent of stipulating what colour of upholstery you want it finished in.

Tackling the enemy

Like the majority of classic cars, Alfa saloons are susceptible to corrosion in fairly specific areas. Predictably, these centre on the rear wheelarches - which are double-skinned - the inner wheelarches, inner and outer sills, door tops on the Supers of the '60s and '70s where the chrome stainless steel trim goes around, the door bottoms, the rims of bonnet and boot and spare

When mud gets trapped behind the headlight rim and within the wheelarch, as in this Giulia Super, corrosion eventually sets in.

wheel well. In short, anywhere that gathers mud or traps water is vulnerable. For instance, mud collects in the valance below the radiator, in the door bottoms if drainage channels get blocked, and in the leading edge of bonnet and trailing edge of bootlid. Condensation within the cabin translates into rust in unseen areas, like the base of the A- and C-pillars. Other problem points can be where trim like grilles, bumper brackets, headlights or fuel filler are attached. Apart from jacking points, the floorpans don't appear to be too bad in the saloons, unlike Spiders, which are more susceptible to leaks. However, the saloon's front panel can be a problem - that section incorporating the valance

between the wings and below the radiator, and ahead of that all-important front-of-chassis crossmember.

Assuming we're dealing with a full restoration rather than a cosmetic job, or localised accident damage repair, the restorer's first act is to strip the car completely. Off comes all the trim, bumpers, hub caps and light clusters,

171

In some cases the steel is virtually intact, and simply needs stripping back completely before being re-painted.

Below left - The front panel of the Giulia Super has been removed, and remedial work carried out on the inner panels before a new front section is fitted.

but corrosion is another matter. Here, the next question to be addressed is whether to patch or panel. For example, if the spare wheel well has only a small hole in it, the easiest solution would be to cut out the rotten area around the hole and weld in a metal patch. But if it bears more relation to a colander, the whole section should be cut out and a complete new panel welded in. The skilled restorer can construct repair patches with contours corresponding to the original section he's replacing.

This philosophy is broadly true for the whole car. The thing about the Giulia saloon is its extraordinary collection of flutes and scallops along roofline and wings, and if these had rusted through at any point it would be a daunting job for the restorer to get them absolutely right by patching and filling. That's another argument in favour of new panels for old. In any case, it's often quicker and easier to fit a replacement wing than it is to cut out small sections of it and replace them with patches. These new replacement panels are readily available from only a small number of specialists, who include Chris Sweetapple's Highwood Motor Company, which charges about £200 for a pattern front panel. Of course, there's the well-known Italian Connection, E.B. Spares, which will supply parts and panels for virtually all classic Alfas. Holland in general is a good source for Giulia saloons, because the cars are so popular there. When living in Amsterdam in 1991, I came across a small operation in the north of the city that specialised in

followed by doors and window glass. Again, depending on the extent of the work, the seats, floor coverings, door rubbers and all interior trim are removed, and most items are labelled to ensure they don't go astray. An exception can be the dashboard, which can be masked to avoid disturbing the electrics, assuming there is no rust lurking in the scuttle behind it. It's all according to the nature of the job whether it stays put or not. Once stripped out, the vehicle is likely to be

spending the next few months on axle stands, and the wheels will more than likely be sent off for shot-blasting before returning for respraying. If period Cromodoras, Campagnolo or Minilites are to be fitted, specialists can do wonders with corroded alloys these days. If the car needs to be moved around the workshop in the meantime, a set of mule wheels is fitted.

Dents can be beaten out and the area made good by dint of careful filling and reprofiling. Dents are one thing

What seems like sheer negligence was actually an act of vandalism when this once proud 2600 Berlina was parked in a farmer's field awaiting restoration.

Below right - Once considered worth keeping just for spares, classics like this 2600 Berlina have become sufficiently rare and valuable to be worth resurrecting no matter what.

restoring nothing else but Giulia Supers. Seeing a group of these cars in the road outside nearly caused me to have an accident of a personal nature. I reeled into the workshop and quizzed the proprietor about his business. It turned out he was bringing them up from Italy, doing whatever needed to be done, and turning them around for the equivalent of about £5000.

While period Alfas with separate chassis such as the pre-war 6C 2300 B will require more heavy-duty specialist work in that department, models produced since the 1900 conform to the fallibilities and structural needs of unit-construction body-chassis units. The sills have much to do with maintaining the integrity of a monocoque chassis, and it is here that Mike Spenceley starts off his guided tour of the Giulia Super restoration. He begins by removing the outer section, which runs from the back of the front wheelarch to the start of the rear wing, conforming to the area immediately below the doors. That's the easy part. To access the inner sill, it's necessary to cut off the lower section of the front wing to the rear of the wheelarch. Having taken that away, you can get at the leading edge of the inner sill. It may be that the internal panels under the front wing are in need of remedial action anyway, since they are in the direct line of fire for mud collection. Depending on condition, the inner sill is either replaced or repaired. Having got as far as accessing it, complete replacement seems the sensible course of treatment here. When it comes to reassembly, Mike administers copious amounts of anti-

corrosive paint before replacing the outer sill panel. This is a type of pigment that's commonly used for painting steel bridges, and it contains zinc, a metal that isn't destroyed by heat when the outer sill - or any other inner panel - is welded on. The lower section of the front wing that was removed in order to perform the operation is refitted and, since this is available as a repair panel, it makes sense to put a new bit on. The inner wheelarch panels that carry the suspension pick-up points may be perforated in places,

and may withstand patching, which is less of a chore than cutting out to make way for a whole new replacement. Generally, of course, outer panels are much easier to replace than inner ones.

Spray something simple

According to Mike, the days of cellulose paint are long gone, with tough, resin-based two-pack having been the order of the day for the last decade. This, of course, requires a different environmental regime and for the painter to be fully protected with an

The sill (rocker) structure of 105-series cars was such that any damp which had accumulated would eventually result in rust, eating away not only the outer layer but the inner skin as well.

Above - A Giulia Super in the throes of restoration at MGS Coachworks. It's had the outer sill fitted and next a repair section will be welded to the lower front wing. Regardless of condition, this section has to be removed in order to access the inner sill.

This is what the inner sill of a Giulia Super looks like. In this case all that was needed was zinc-based anti-rust treatment before the new outer sill was fitted over the top.

The outer sill has been replaced on this Giulia Super. This shot makes clear that the car has to be gutted before a thorough restoration can begin, with all interior trim, window glass, doors and closure panels removed to access hidden areas. The car is locked in a jig or sits on axle stands throughout most of the operation.

Below - One of the places that can be a nightmare is the Giulia Super's spare wheel well, but in this case it's only suffering from minor surface rust.

Below - The same boot (trunk) floor after it's been painted. The thickness of three layers of two-pack paint is the equivalent of over 20 coats of cellulose, something the 1960s cars would never have got, especially in the boot space.

air-fed mask because of isocyanate poisons in the pigment. It should also be used in a proper low-bake oven. All paints are of a lower solvent content these days, and body shops have to keep a stringent record of solvents used.

Mike sprays his cars with two coats of etch primer, then a high-build primer is applied on top of the etch primer, and the two coats bond together. The surface is then flatted down and any imperfections made good. The car is re-primed with another two or three coats of primer, depending on the level of rectification that's taken place. If the car is being painted with a

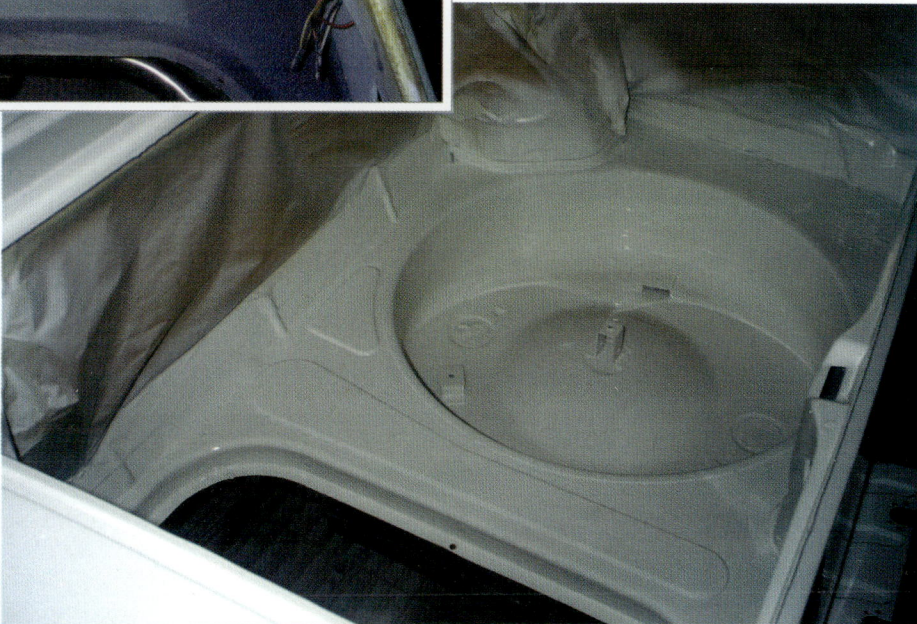

Attention to detail includes painting the suspension components, as well as treating the inner wheelarch with stone-chip protection and paint. This is as good a time as any to fit new components like these Koni Classic dampers.

metallic paint, it's rubbed down with a finer wet and dry paper, and if it's to be a solid colour it will be finished with a slightly coarser paper. This is because the metallic paint acts as a base coat, which is then lacquered, and in the first place it's a thinner consistency than the solid colour.

Mike - or his staff - then apply three layers of two-pack colour coat, assuming the car is to be finished in a solid hue, as opposed to metallic. The thickness of these three coats is the equivalent of no less than 21 coats of cellulose, which seems impressive, but it's actually a fallacy that more paint on the car is better, because it can chip more easily. What is important is the way it's applied, and that's where the specialist painter excels.

To obtain a metallic finish, three colour coats are followed by two coats of lacquer. Pigments dry extremely quickly, and after something like eight minutes, another coat of paint or lacquer can be applied.

When the painting operation is complete, all inner sections are injected with cavity wax, which is the same material used in construction of modern Alfas, Fiats, VWs or whatever. Waxoyl has become largely a thing of the past as far as restorers are concerned because it runs when hot and

needs to be re-applied every so often. Cavity wax dries to a film on the inner box sections, and thus is permanent.

Restoration requirements can vary according to the age of the vehicle. If you take the Giulia range, you sometimes find that later cars are more rust-prone than early '60s models, because of the use of slightly thicker gauge steel in the earlier cars, and also metal less compromised or adulterated by recycling, which is fundamentally what did for the Alfetta, Giulietta

and Alfasud. But time rolls on, and the youngest 105-series Giulia Super is going to be over 25 years old now. Unless it was built in South Africa or lived in Southern Italy, the chances are it'll have already had some remedial work done, which, with luck won't have been a bodge-up. Being more high profile and overtly desirable, the Coupés were the first to feel the burning and cutting of the restorer's blowtorch, and it's only more recently that the more common saloons have

It's rare to find close to pristine original examples of classic cars these days. But Richard Everton was fortunate enough to come across this T.I. Super in Provence, southern France, which became a potential concours champion with a little specialist remedial work.

been deemed worthy of restoration. There is thus a better chance that they won't have been mucked about with to the extent that some unfortunate Spiders and Coupés were in the first half of the 1980s. The Berlinas weren't worth much then - or now, relative to their specification and grin-factor - so owners were rarely motivated to splash out on a restoration.

In the past, some Alfa owners bought cars cheaply and expected to run them on a shoestring. But times change and they, along with the classic car speculators, have largely disappeared from the scene, or at least grown wiser and wealthier. Now, it seems there are more genuine enthusiasts, who genuinely appreciate the cars and have the wherewithal to keep them in top condition. These people are prepared to fork out for new panels rather than have a restorer cutting and splicing in patches of new metal in order to achieve the polished result. To an extent, it's a trend that MG restorers brought to the attention of the wider world with MGBs re-shelled with Heritage bodies in the early 1990s, and that's excellent news for Giulia T.I. and Super fans, as well as for posterity.

AROC Chairman Ed McDonough found this Giulietta T.I. in a cave in Naples, having had only a little rust treatment before being hidden away for 28 years. Ed plans to restore the car in the racing colours of the legendary Pedro and Ricardo Rodriguez, and do some historic rallying with it.

Cave dweller

Some extraordinary survivors turn up from time to time; cars that we've all dreamed about discovering hidden away in the back of a barn, a notion that largely passed into mythology in the late 1980s.

But they do turn up from time to time, and a couple of years ago, AROC Chairman Ed McDonough was lucky

enough to hear of a 1958 Giulietta 1300 T.I. that was languishing in a local authority reclamation cave in Naples. He caught a train to Italy and went to see it, discovering that it was in fairly good original condition, and because the seats had been covered over, the interior was immaculate. In spite of having spent 28 years in this cave, it even started first time! It had appar-

Ed McDonough's Giulietta T.I. is masked up with a coat of primer on. The car was so good that only one sill needed replacing and the brightwork re-chroming.

ently had a new clutch just before its incarceration, and had covered only about 30k miles in total. Ed did a deal (around £1100) and shipped the car to the UK via Dublin. By the time you read this, he'll have done any necessary bodywork restoration himself - 'The first time I've done anything like that,' he told me. 'It was parked near to one of the cave walls, so that side of the car did need some attention.' As well as an engine and suspension rebuild, Ed plans to paint the Giulietta in the racing colours of the celebrated Rodriguez brothers, Ricardo and Pedro. It all gets a bit convoluted here, but on a recent trip to Mexico, Ed learned from the late Moises Solana's brother that the Scuderia Rodriguez ran two Giulietta T.I.s in 1957 and 1958, and that one Conrero-engined car certainly was painted dark blue - Ed even met the guy who painted it. Back then, Pedro was aged 17 and Ricardo was only 15, yet they were already good enough to beat even veterans of the *Carrera Panamericana* with their Giulietta T.I.S. The Scuderia Rodriguez logo was suitably combative in concept, featuring an eagle in flight, with a cactus branch in its talons and a snake in its beak. Ed McDonough, who's writing a book about the Rodriguez brothers, proposed to have his own car finished in the *Scuderia Rodriguez* colours, prepared to Appen-

dix K specification for Historic Production Touring Cars and indulge in a little light competition with it. Lucky man. And if you're wondering, Moises Solana drove for the Cooper, Lotus and Centro Sud teams in the 1960s.

Holy bodies
Jumping a generation, the Alfettas and Giuliettas of the 1970s and early '80s are approaching classic status. The good news is that they were mainly pretty reliable and excellently engineered, but the bad news is that the bodies were flawed. 'Deeply

Models from the 1970s, like this Giulietta, were probably more prone to rusting than earlier cars because of the widespread use of corrupted steel.

178

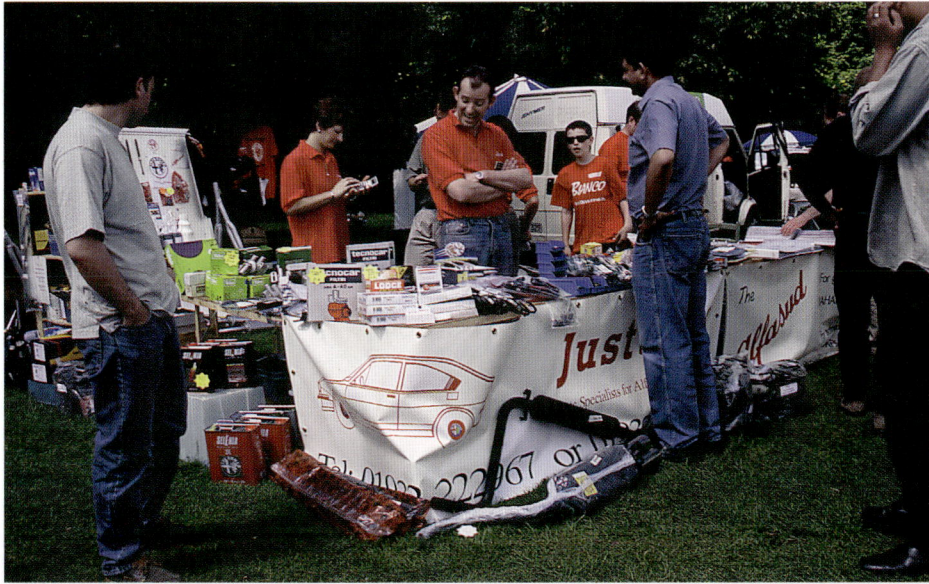

Owners of classic and not-so-classic Alfa Romeos can source new and used parts for their cars from dedicated specialists such as Just Suds, whose stall, doing a roaring trade at the UK National Alfa day, is staffed by Ian and Ann Brookfield and Joanne Eastwood.

depressing to own,' says Mike Spenceley, reflecting on his previous tenure of a 116-series Giulietta, 'because they rotted away around you. They were so good to drive, but the only items that didn't succumb were the

The Alfa 75-based SZ may have been a £40k supercar, created in the enlightened times of the late 1980s when people knew about the causes and effects of corrosion, but the combination of steel and aluminium which framed the car's plastic roof still reacted against each other to cause corrosion.

Mike Spenceley's own Giulia Super is testimony to what a full restoration can do for a car. It's fair to say that the car is probably better than when new, because all the underbody areas will have been properly treated, an aspect that received scant attention when these models were originally made.

doors - those they seem to have protected quite well.' He's restored only one Alfetta saloon to date, and it seems that there's not the interest or affection yet for this model that surrounds the Giulia Super. Similarly, Alfasuds are more numerous, and a comprehensive restoration is unviable in harsh financial terms because the cars just aren't worth that much, however affectionately people regard them.

From my own experience, I can tell you that there wasn't much change out of £10k when the GTV6 was restored back in 1991. One day it may be worth that kind of investment, though.

There were also mechanical shortcomings that affected the Alfetta and Giulietta ranges, including the GTV coupés, which need to be addressed if ownership is contemplated. These included a tendency for wheel bearings to wear, head gaskets to blow or valves to burn out, and front wheel bearings became noisy if they were over-tightened, while the three rubber doughnut couplings in the driveshaft were prone to splitting. On the corrosion front, the clips and screws that attached external trim broke the surrounding paint finish and rust took hold. Mud would build up on a ledge at the rear of the inner front wheelarches and eat away around the scuttle and base of the A-pillar.

The 75 is very popular in club circles, but it's getting a bit long in the tooth now. It started life with a bit of a head start over the 1970s models in that Alfa already knew how crippling the corrosion legacy was in sales terms, and restoration specialists had shown how it could be avoided, or at least kept at bay. Prevention is better than cure, as we know, and in order to stave off corrosion, the 75 received special attention in preventative treatment on the production line. Assembly-line workers at Arese injected wax into box sections and cavities that could harbour rust, which has made the cars last much better. Mine, a 1988 car, was around ten years old - and 150,000 miles down the road - when it first showed signs of beginning to rot where the rear light clusters meet the rear wings. Paint was also starting to bubble just to the rear of the point where the bulging front wheelarches of the Veloce body kit merge with the wings.

And the man who bought it from me last year, bless him, found a hole in the floorpan that I didn't know about.

Another corrosive curiosity relating to a derivative of the 75, the bulldog-like SZ coupé, was thrown up by Mike Spenceley. He was doing a restoration job on an SZ for none other than Les Dufty, the widely known and respected Bristolian Alfa tuner and proprietor of Automeo. Les discovered that the media mix of aluminium, plastic and steel which combined to create the car's distinctive roof panel were not happy neighbours, and corrosion was beginning to appear. That's a rather exceptional case and, in general, Alfas have shrugged off the cloud that hung over them during the 1980s; some fine cars are being put back on the roads thanks to diligent restorations and caring owners. As you'll notice from some of the illustrations, Mike had recently begun work on a 1971 Giulia Super that had stood in someone's garden for 17 years, grown 'window boxes' in the process, and yet only required sills and work on the boot floor. I've asked for first refusal when he's finished it!

APPENDIX - ALFA SPECIALISTS & CLUBS

Alfa Romeo Berlina owners and buyers may find the following list of specialists useful. These are not Alfa main dealers, since such franchises seem to swap around, or change hands, quite frequently these days. The companies listed here have all been in the game for some time; however, for a fuller and up to date listing, check with the AROC or their equivalent in your country. Please note that the inclusion of a specialist in the following list is not a guarantee of good service and that neither publisher nor author can accept any responsibility for any problems you may encounter with these independent companies.

1900 Register
Peter Marshall
Mariners
Courtlands Avenue
Esher, Surrey, England.

2600 Register
Roger Monk
Ettrick House
8 Cambridge Road
Hesketh Park
Southport PR9 9NG
England

AC Trofeo Motorsport
Nigel Cottee or Alan Marshall
Unit 11, Churchfeld Court
Top Valley
Bewcastle Road
Nottingham NG5 9PJ
England
Tel: 01159 204211

AFRA SAS
(Rare Alfa Romeo spares, ex-factory or made to order)
Snra. Croci
Via Carducci 36
20019
Settimo Milanese
Italy
Tel: 0039 02 3286111

Alfa II
(Alfa servicing, sales from 1970 onwards, exchange engines)
Ramesh Bharadia
Unit 5
Parr Road
Stanmore HA7 1NL
England
Tel: 0181 951 4100

Alfa Parts Exchange
(Used components, 1950-1994 US-spec cars)
Larry Dickman
2000 National Avenue
Hayward, CA 94545
USA
Tel: 001 510 782 5800

Alfa Ricambi
(New and used parts, reconditioned, engine & suspension upgrades)
Julius T. Mann or Brad L. Bunch
6644 San Fernando Road
Glendale, CA 91201
USA
Tel: 001 8189567933

Alfa Romeo Giulia 105 Register
Stuart Taylor
144 Sussex Way
Cockfosters

Barnet
Herts EN4 0BG
England

Alfa Romeo Giulietta Register
Peter Shaw
Grange Farm House
2, Bedford Road
Willington, Beds MK44 3PS
England

Alfa Romeo Owners' Club
Secretary: Michael Lindsay
97 High Street
Linton
Cambridge CB1 6JT
England
Tel: 01223 894300

Alfa Romeo Owners' Club Shop
Ray Skilling
EM Models
42 Camden Road
Tunbridge Wells
Kent TN1 2QD
England
Tel: 01892 536689

Alfa Servizio
Colin Wing
Unit 79
Bunting Road Industrial Estate
Northampton NN2 6EE
England

Alfa Stop
(Classic Alfa brake, clutch, transmission & exhaust systems specialist, for 1900, 2600, Giulietta etc)
Tony Stevens
PO box 50
Belper
Derbyshire DE56 1AS
England
Tel: 0177 382 2000

Alfarama
Westmoreland Road
London NW9 9RL
England
Tel: 0171 206 2075

Alfashop Ltd
(Importer of spare parts, Alfasud/Alfetta specialist)

Jeremy Wales
Unit 1 Beech Drive
Mile Cross Lane
Norwich NR6 6RN
England
Tel: 01603 426277

Alfatune
(Road, race & rally preparation)
Gus Lambrou
Merton Bank Road
St Helens
Merseyside WA9 1HP
England
Tel: 01744 25499

Arese T.H.O.
(Parts, repairs, tuning)
Ing. C.J. Violier
Leeweg 7
1161AA Zwanenburg
Netherlands
Tel: 0031 543 451906

Autodelta
(Customising specialist)
Jano Djelalian
Forza House
St Leonards Road
London NW10 6ST
England
Tel: 0181 838535

Automeo
(Alfa Romeo servicing, spares, maintenance; carburettor specialist)
Les Dufty
36 Gypsy Patch Lane
Little Stoke
Bristol
England
Tel: 0117 969 5771

Avon Racing
(Race car rental, e.g. 75 Twin Spark, suspension uprating)
Clive Hodgkin
6 Station Road
Uppingham
Leicestershire
England
Tel: 01572 823514

Richard Banks
(Classic Alfa Romeo and Italian car sales)

Oakford, Tiverton
Devon
England
Tel: 01398 351360

Ken Bell
(Service, repairs)
St James's Road
Fleet
Hants GU13 9QH
England
Tel: 01252 629159

Bell and Colvill
(Alfa Romeo specialist)
Bobby Bell or Martin Colvill
Epsom Road
West Horsley
Surrey KT24 6DG
England
Tel: 014865 4671

Benalfa Cars
(Alfa Romeo restorations, engine rebuilds)
Alan Bennett
19 Washington Road
West Wilts Trading Estate
Westbury
Wiltshire
England
Tel: 01373 864333

Black & White Garage
(Former main dealer, sales & service)
Hermitage Road
Cold Ash
Thatcham, Berks
England
Tel: 01635 200444

BLS Automotive
(Dealership, plus race preparation)
Tom Shrubb or Phil Bower
Unit 1 Gt Northern Way
Lincoln LN5 8XF
England

Bridge Alfasport
(Engine & gearbox specialist, 164 service & repairs)
John Brooke
Bridge Garage
Henfield Road, Cowfold
West Susses RH13 8DT
England

The Carburettor Hospital
(UK's largest carburettor stockist, Solex spares from 1933)
Eric Archer or James Cooper
210 Woodgrange Drive
Southend-on-Sea
Essex SS1 2SJ
England

Carsmiths
(105-series restorations, maintenance, replica GTAs built)
Jonathan Smith
Unit 6, Old Station Yard,
Chapel Street, Cawston,
Norwich
Norfolk NR10 4BG
England
Tel: 01603 872568

Classic Alfas
(Service & MoT checks, parts sourcing for Giulietta models)
Ian Williams
Camden Garage
9 Camden Terrace
Weston-super-Mare
Somerset BS23 3DH
England

Classic Car Interiors
(Original pre-1976 Alfa Romeo upholstery)
Chris Adams
Park Farmhouse
Cornworthy, Totnes
Devon TQ9 7ES
England
Tel: 01803 732454

Cloverleaf Transmissions
(Mechanical overhauls, especially transmissions)
Charlie Skinner
70 Minet Avenue
Harleston
London W10 8AP
England
Tel: 0181 961 2355

T.A. & J.M. Coburn
(Alfa Romeo soft-top and upholstery specialist)
Widhill House
Blunsdon
Swindon, Wilts

England
Tel: 01793 721501

C.P. Garage Services
(Servicing & parts)
Euan Colbron
Unit 3 Blauniefield Industrial Estate
Dundee DD4 8UT
Scotland
Tel: 01382 731479

Cubleys of Ainsdale
(Alfa dealer, new & used spares)
Mike Haliday
609 Liverpool Road, Southport
Merseyside PR8 3NG
England

Bob Dove Motorsport
(Race preparation, bits & pieces)
71 Celeborn Street,
South Woodham Ferrers
Essex CM3 7AF
England
Tel: 01245 328278

Richard Drake Motors
(Alfa Romeo servicing, engine rebuilds, race preparation)
Unit 2, Renson Close
Beech Drive
Mile Cross Lane
Norwich NR6 6RH
England
Tel: 01603 406050

DRH Developments
(Race preparation, mechanical & electrical work)
Dave Hood
20 Daventry Road
Dunchurch
Warwickshire CV22 6NS
England
Tel: 01788 815936

EB Spares (The Italian Connection)
(Parts for 105-series Giulias, plus Alfettas & 75s)
David Edgington or Kevin Abigail
31 Link Road
Westbury Trading Estate
Westbury
Wilts BA13 4JB
England
Tel: 01373 823856

Jim Evans
(Race preparation of 105-series engines & suspension)
Scagglethorpe Manor
Malton
North Yorkshire YO17 8DT
England
Tel: 01944 758909

Excellent Bodyshop Ltd
(Restoration & repair of bodywork)
Nino di Luca
8 Johnson's Way, Park Royal
London NW10 7QB
England
Tel: 0181 453 0282

Frenitalia Tarox
(UK importer of Tarox brake system)
R. Cappucci
Unit 9, Taylor Court
Carr's Industrial Estate
Haslingen BB4 5LA
England
Tel: 01706 222872

Gatwick Alfa
(Spares & repairs, 1950s to 1990s, plus race preparation)
Mike Buckler
Rusper Garage
High Street, Rusper,
Horsham
West Sussex RH12 4PX
England
Tel: 01293 871155

John Goodchild Motor Engineers
(Repairs, servicing, maintenance, especially 105-series cars)
112 Turnpike Link
Croydon CR0 5NY
England
Tel: 0181 680 2120

P. D. Gough & Co
(Hand-built stainless steel exhaust systems)
The Old Foundry, Common Lane
Whatnall
Nottingham NG16 1HD
England
Tel: 01159 382241

Gran Turismo Engineering
(Classic Alfa Romeo specialist, sales,

race preparation)
Simon Whiting
Station Avenue
Kew Gardens
Surrey
England
Tel 0171 460 0007

Harvey Bailey Engineering Ltd
(Suspension tuning & set up for 105-series, Alfetta, 75)
Anne Harvey Bailey
Ladycroft Farm
Kniveton, Ashbourne
Derby DE6 1JH
England
Tel: 01335 346419

High Performance Restorations
(Bead-blasting & powder-coating a speciality)
Robert Thompson
Oak Farm, Wilcot Lane
Nesscliff, Shrewsbury
Shropshire SY4 1DB
England
Tel: 01743 741592

The Highwood Motor Company
(Alfa Romeo replacement panels specialist, workshop manuals on CD ROM)
Chris Sweetapple
137, Bishopston Road
Swansea SA3 3EX
Wales
Tel: 01792 234314

Peter E. Hilliard & Son
(Servicing & repairs, all Alfas from 1963)
41 High Street, Penge
London SE20 7HJ
England
Tel: 0181 778 5755

Nick Humphrey Motorsport
(Race preparation of Boxer engines)
Coppice Corner
3 Castlemaine Drive
Hinckley
Leicester LE10 1RY
England
Tel: 01455 618575

The Italian Car Co
(Alfasud & 33 specialist)
Lawrence Garnett
Unit 2, Carlton Farm
Beehive Lane
Chelmsford
Esex CM2 8RJ
England
Tel: 01245 494755

Italian Miniatures
(Alfa Romeo models)
Richard Crompton
39 Penncricket Lane
Oldbury
Warley
West Midlands B68 8LX
England
Tel: 0121 559 6611

Just Suds
(Alfasuds and Sprints)
Ian Brookfield
24 Dallington Close
Hersham
Surrey KT12 4JG
England
Tel: 01932 222967

Just 33s
Steve Wrighting
Unit B2 Greenwood Court
Veasey Close
Attleborough Fields
Nuneaton CV11 6RT
England
Tel: 01203 370919

K & L Autos
(Mobile servicing of classic Alfas)
Keith Waite
Golders Green
London N5
England
Tel: 0181 458 3879 or 0585 655503

Alwyn Kershaw
(Alfa dealers, special tuning)
Peter Colley
Langford Garage
Helmsley
York YO4 1NF
England
Tel: 01759 373399

Kinghams of Croydon
(Former Alfa main dealer, sales, servicing)
Keith Kingham
39/41 South End
Croydon CR0 1BE
England
Tel: 0181 680 0922

Lombarda Carriage Company
(Alfa spares and sales)
Colin Fallon or Mark Wakeford
3-10 Railway Mews
London W10 6HN
England
Tel: 0171 243 0638

London Stainless
(Stainless-steel exhaust systems for virtually all post-war Alfas, mail order)
Giles Beaumont or Paul Goddard
251 Queenstown Road
Battersea, London SW8 3NP
England
Tel: 0171 622 2120

Lyles of Newcastle Ltd
(Sales, parts & service)
Jason Sanderson or Ian Dodsworth
Milano House, West Road,
Newcastle-upon-Tyne NE15 6PQ
England
Tel: 0191 2730700

Mangoletsi
(Alfa sales, service, parts)
Knutsford
Cheshire
England
Tel: 01565 722899

MGS Coachworks
(Alfa restoration specialist)
Mike Spenceley
2 Foxley Hill Road
Purley
Surrey CR 8 2HB
England
Tel: 0181 645 0555

Monza Sport
(Alfa sales, old and new)
John Griffiths
High Street
Rusper
West Sussex RH12 4PX

England
Tel: 01293 871900

MP Racing
(Performance tuning of older Alfas)
Piero Pesaro
Unit C, Davis Road
Chessington
Surrey KT9 1TT
England
Tel: 0181 974 1749

Richard Norris
(Classic Alfa Romeo spares specialist)
44A The Gardens
East Dulwich
London SE22 9QQ
England
Tel: 0181 299 2929

Brian Pillans
(New and used parts for post-1965 Alfas)
129 Maxwell Drive
Glasgow GH1 5AE
Scotland
Tel: 0141 3313424

Jamie Porter
(Alfa servicing, spares, rolling road, maintenance)
Rowland House
Lower Gower Road
Royston, Herts
England
Tel: 01920 822987

Premier Garage
(Service, repairs, tuning etc)
Rob Kirby
High Street
Newport
Essex CB11 3PE
England
Tel: 01799 541041

Prinz Bredevoort
(Parts for 105-series cars to 1990)
Henk Prinz
Kleine Gracht 10
7126 AW Bredevoort
Netherlands

R. Proietti Ltd
(Maintenance of Alfas of all ages)
Stef or Bruno

2 Blundell Street
London N7 9BJ
England
Tel: 0171 607 0798

RacingBrand Ltd
(Performance and styling upgrades, bodykits, suspension, traction control etc)
David Moore
Pangbourne Lodge
Pangbourne
Berks. RG8 7AZ
England
Tel 01189 844175

Ramponi Rockell
(Alfa Romeo sales, servicing)
Alex Jenkins
30-31 Lancaster Mews
London W2 3QE
England
Tel: 0171 262 7383

Rossi Engineering
(Race preparation, restoration, maintenance)
Rob Giordanelli
Sunbury on Thames
Surrey
England
Tel: 01932 786819

Schwaben Garage
(Specialising in high performance engine conversions, especially 75s)
Steve Fielding
Unit 10, Elmcott Farm
Great North Road
Biggleswade
Beds SG18 9BE
England
Tel: 01767 683681

Spider's Web
(Rebuilds & restorations)
Roger Longmate
Westgate Street
Hilborough,
Thetford
Norfolk IP26 5BN
England
Tel: 01760 756229

Sunnyside Garage
(Mechanical & electrical repairs)

David Lai
Unit L5, Chadwell Heath Industrial Park
Kemp Road
Dagenham
Essex RM8 1SL
England

Julius Thurgood
(Classic Alfa sales, competition car specialist)
Broomfield Farm
Coleshill Road
Bentley
Warwicks CV9 2JS
England
Tel: 01827 720361

Timeless Motor Company
(Overhauls, repairs, restorations, especially 105-series)
John Timpany
Pooles Lane
Highwood
Chelmsford
Essex
England
Tel: 01245 248008

Touring Superleggera
(Classic Alfa restoration, race preparation)
Andrew Thorogood
172 Clapham Park Road
London SW4 7DU
England
Tel: 0171 720 8616

Fa. Tuynder C.V.
(Parts by mail order)
J. Tuynder
Westvlietweg 40
2267 AB Leidschendam
Netherlands
Tel: 0031 70 3874403

Veloce Cars
(Secondhand Alfa sales, spares)
Romford
Essex
England
Tel: 0181 551 0644

Veloce Sport
(Car sales and parts, from 1960s to 1990s)

Adam Andrews
Pine Warren Boston Road
Heckington
Nr Sleaford
Lincs NG34 9JF
England
Tel: 01529 469035

W.A.D. Auto Developments
(Service, maintenance, parts, etc)
Willie Dick
3, Waterside Industrial Estate

Ettingshall
Wolverhampton WV2
England
Tel: 01902 401711

Westbury Insurance Serives
(Alfa specialsist, sponsor of AROC race series)
23 Castle Street
High Wycombe
Bucks HP13 6RU
England

Tel: 01494 444574

Westune Auto Services
(Servicing, tuning, etc)
Peter or Simon West
Marsh Street
Horwich
Bolton
Lancs BL6 7TA
England
Tel: 01204 697535

AlfaRomeo

INDEX

Abarth: 51,137
Abthorpe, Mike: 60
Aisin automatic transmission
   Q-system: 157
Alboreto, Michele: 143
Alen, Markku: 143
Alessio, Carrozzeria: 17
Alfa Corse: 32,106,111,125,128,
   135-37,140,143
Alfa Romeo vehicles (in chronological
order):
   24HP: 8,12-14
   12HP: 12-13
   15/20HP: 8,13-14
   14/16HP: 8,13
   20/30HP: 8,14-15
   20/30 ES Sport: 15,17
   30-50HP G1: 16-17
   40-60HP Aerodinamica: 13,14,25
   R.L. Normale/Super Sport: 9,17-20
   R.M. Normale/Sport: 18-20
   P2: 18,20,28
   6C 1500: 8,20-21,24-25
   T500 Autobus: 21
   6C 1750 Turismo/Gran
   Turismo: 21-24,26,28,78
   6C 1900 Gran Turismo: 25-26
   6C 2300/B/Pescara/Mille Miglia,
     etc: 7,9,26-32,34-36,173
   T85 Autobus & truck: 29,35
   T50 truck: 29
   T110 coach: 29
   8C 2300: 25,28
   P3: 28
   8C 2900/B: 30,32-33,35
   12C Tipo C36: 30
   Tipo 158/159 Alfetta: 37,42,91,151
   6C 2500/B/Sport, etc: 7,34-42,
     99,153
   Gazzella prototype: 38
   Tipo 800 Autocarro: 35
   A110 Autobus: 37

   A430 Autobus/truck: 37
   6C 2500 Villa d'Este: 37-39,42,151
   6C 2500 Super Sport Freccia d'Oro:
     7,38-40,42,44,62,151
   6C 3000 CM: 47
   1900 Berlina, Normale/T.I./Super:
     6,8,42-47,57-58,60,62,151,173
   1900 Super Sprint: 42,44,46-47
   1900 Matta AR51/52 Jeep/
     Giardinetta: 42,47
   Tipo 455 truck: 42
   Tipo 900 truck: 42
   Romeo F12 van: 46,56,78
   B.A.T. cars: 14,47
   Giulietta Berlina Normale/T.I.,
     Promiscua, etc: 6,7,11,48-55,57,
     59,62,65,82,85,91,118,151,177-
     78
   Giulietta Sprint/Veloce: 6,20,46,48,
     50-52,54,69,151
   Giulietta Spider: 6,50-52,69,176
   Giulietta/Giulia Sprint Speciale (SS):
     54,58,66,69-70
   Giulietta SZ/Coda Tronca: 4,9,52,54,
     58,63-64,148,151
   Giulietta SVZ: 4,52,54
   AR Renault Dauphine/Ondine: 55-56
   AR Renault R4: 56
   Tipo 1000 truck: 56
   Sicar/Turbocar coaches: 56
   2000 Spider: 58
   2000 Berlina: 57-58,62,99
   2000 Sprint: 58
   Tipo 103 prototype: 55
   2600 Berlina: 4,58-62,99,173
   2600 Spider: 59-61
   2600 Sprint/SZ: 4,59-61
   Giulia 101-series Sprint: 48,69
   Giulia 1600 T.I.: 9,11,56,59,62,64-
     66,71,78,91
   Giulia T.I. Super: 10,65-71,80,170,
     177

Giulia 1300 T.I.: 10,11,71-73,82
Giulia 1600 Super/S/Nuova/
Giardinetta, etc: 4,10-11,54,64,68-
69,73-76,78,80-82,85,92-93,118,
148,152,155,170-77,179-80
Giulia Sprint GT/GTV: 58,62,69-74,
79-81,84,88,93,148,170,176
Giulia Sprint GTA & derivatives:
9,10, 54,68-71,73-74,80-82,86,88,
93,108,170-71
Giulia GT-Z/TZ2: 7,54,63,148
1600 Duetto/Spider: 62,69-71,82,
151-52,171
Montreal: 86,109,142
1750 Berlina: 6,10,78-82,162
2000 Berlina: 11,72-73,78,80-83,89,
91-92
Junior Z: 71
Tipo 33 sports prototype: 63,86,109,
114
Tipo A12 truck: 86
Tipo A38 4x4 pickup: 86
Alfasud & variants: 6,9,56,83-91,94,
113-18,120,144,176,179
Giardinetta: 86,90,118
Sprint: 86-90
Arna: 56,113-14,117
Alfetta: 91-100,103,107,132,149,
170,176,178-79
Giulietta: 95-99,103-104,107,115,
176,178-79
Alfetta GTV/GTV6: 89-90,93,95,97-
98,101,103,105-106,109,149,
170,179
Alfa 6: 98-103
Alfa 90: 6,102,103,109
Alfa 33: 11,83,90,107,114-121,128,
130,144,147-49
Giardinetta /Sport Wagon: 107,117-
20,145
Alfa 75: 9,11,90,95,98,103-12,115-
16,120,125,128-29,131-33,135,
141,162,179-80
Alfa 164: 5,56-57,112,114,120-31,
133,137,165-67
SZ/ES30: 110-12,179-80
Alfa 155: 4,5,10,112,115,129,130-43,
145,147,149,153-155,161-62,164
Spider: 133,149,161,167
GTV: 130,133,149,161,167
Alfa 145: 144-150,152-53,164,166
Alfa 146: 4,147-150
Alfa 156: 4,6,11,133-34,151-67,169
Sports Wagon: 162
Alfa 166: 10,57,165-69
Alfa 147: 150

Alfa Romeo factories & sites:
Arese: 8,14,18,25,32,35,56,62,63,70,
73,74,79,92,99,101,123,129,135,
179
Avellino/Pratola Serra: 113,147
Balocco test track: 8,62
Pomigliano d'Arco: 35,37,56,83,
84,116,118
Portello: 8,12-15,20,23-26,29,30,35-
38,40,42-44,46,48-50,54,56,63
Alfa Romeo personnel:
Anderloni, Carlo Bianchi: 99
Bazzi, Luigi: 18
Busso, Giuseppe: 44
Castelli, Pierguido: 135
Chirico, Domenico: 84
Chiti, Carlo: 67
Chizzola, Lodovico: 67
Colucci, Ivo: 44,62
Cressoni, Ermano: 123
Ferrari, Enzo: 8,16-18,35
Fucito, Edoardo: 8,21
Fusaro, Piero: 122
Gallo, Pasquale: 21,36,56
Ghidella, Vittorio: 111,122
Gianferrari, Prospero: 21
Gobbato, Ugo: 26,36
Guidotti, Giambattista: 44
Hruska, Dr Rudolf: 83,98,114
Jano, Vittorio: 7,18,21,25,30
Limone, Sergio: 135
Luraghi, Giuseppe: 56,67,84
Masoni, Edo: 109
Merosi: Giuseppe: 8,12,16-18
Moneta, Antonio: 62
Pavese, Giorgio: 62
Pianta, Giorgio: 106,135-36
Piccione, Alessandro: 131-32
Ramponi, Giulio: 17,21
Ricart, Wilfredo: 30,36,38
Rimini, Giorgio: 8,21
Romeo, Nicola: 8,14,15,17-18,21
Sala, Silvio: 62-63
Scarnati, Giuseppe: 62-63
Sfondrini, Renzo: 62
Sivocci, Ugo: 17
Russo, Nini: 135-36
Sanesi, Consalvo: 38,44,46
Santoni, Antonio: 8
Satta Puliga, Orazio: 36,50,62
Scarnatti, Giuseppe: 50
Travaglio, Gianclaudio: 165-66
Alfa Romeo Owners' Club: 4,102,107,177
AROC Championship race series: 87,
89,97-98,105-108,114,120-21,
127-28

National Alfa Day: 4,66,162,170,179
156 Challenge: 163
Alfashop: 72
Alfatech: 105
Alpine Rally: 46,54,81
Amthor, Stig: 143
Anderloni, Felice Bianchi: 21,31
Andretti, Michael: 106
Andruet, Jean-Claude: 93
Armstrong Siddeley: 21
Assen circuit: 10,76
Aston Martin: 122
Audi:
A4: 11,141
A6: 167
Quattro: 109,118
Autodelta Racing Team: 8,9,67-71,73-74,
81,106
Auto Italia Championship: 90
Automeo: 180
Avanzo, Maria Antonietta: 17
Avio turbocharger: 98
Avon Motor Tour of Britain: 92
Avus circuit: 143

Baggioli: 67
Baghetti, Giancarlo: 68
Balbo, Carrozzeria: 35
Baljon, Bart: 52
Banca Italiana di Sconto: 17
Banca Nazionale di Credito: 17
Banca Nazionale di Sconto: 26
Barfoot, Mike: 94
Barilla, Paulo: 106
Bartels, Michael: 143
Bartlett, Kevin: 68
Benalfa: 170
Benbow, Dave: 139
Bennett, Alan: 170
Berlin Motor Show: 34-36
Bertaggia, Enrico: 140
Bertone, Nuccio/Carrozzeria: 6,47,58-
60,70,78,122
Betti, Bruno: 100
Bianchi Automobiles: 8
Bianchi, Lucien: 81
Biela, Frank: 141
Birmingham Motor Show (NEC): 166
Biscione: 67
B.L.S. Automotive: 107,108
BMW:
1800Ti: 9,70
2002 Tii: 10,74
3-series: 11,120,129,162,164-65
5-series: 11,125,127,167
Boano, Mario/Carrozzeria: 46-48,145

Boneschi, Carrozzeria: 35-38,40,44,47
Booker, Keith: 75-76,96
Borgward: 10
Bosch:
    ABS: 125,133-34,149
    distributors: 124
    fuel injection/engine management
      system: 95,101,104,108-109,120,
        124-25,132,147,156
    common rail diesel system: 158
    headlights: 24
    starter motor: 20
Bosato: 67
British Touring Car Championship
(BTCC): 135-37,139-40,142,164
Brabham: 128
Brakes: 45,49,71,73,75,85,92,104,147,
    160,169
    ATE: 9
    Brembo: 136,143,165
    Dunlop: 9,67
Brands Hatch circuit: 10,74,94,107,
    121,127,140
Brescia Grand Prix: 17
Brittan, John: 4,53
Brookfield, Ian & Ann: 4,90,179
Buehrig, Gordon: 30
Busi, Giambattista: 140
Bussinello, Roberto: 67,70

Caddell, Laurie: 4
Cadwell Park circuit: 89,106
Campari, Giuseppe: 21,28
Carburettors: 74,117
    Dell'Orto: 69,81-82,92,99
    Memini: 25
    Solex: 21,44-46,50-51,54,58-59,64,
        72,81-82,84-85
    SU: 44
    Weber: 33,44-45,51,66,71,74,85
    Zenith: 17,20
Carrera Panamericana: 38,46,178
CarWeek magazine: 132,139
Castagna, Ercole/Carrozzeria: 14,17-18,
    30,34-35,37,41,47
Castle Combe circuit: 98
Castle, James: 170
Catt, Ian: 4,70
Cazzago, Gherardo: 140
Centro Internazionale Guida Sicura
(track days): 164
Chapman, Colin: 10
Cheffings, Jane: 127
Chiesa, Filippo di: 6
Citroën:
    GS: 84,87-88,147

Cleland, John: 139
Colli, Carrozzeria: 6,44,47,54-55,118,
    171
Columbo Borriane cams: 69
Comprex supercharger: 118
Conrero, Virgilio: 67-68,178
Consten, Bernard: 54
Cord 810: 30
Couceiro, Pedro: 143
Crumpler, Shane: 114
Cudini, Alain: 95
Curtis, Andy: 121

Dack, John: 110
Dampers:
    Bilstein: 136
    Intrax: 70
    Koni: 73,108,176
    Spax: 109
Danner, Christian: 143
Darracq, Alexandre: 8
De Adamich, Andrea: 10,54,67-
68,70,164
De Adamich, Gordon: 140,164
De Lageneste, Roger: 54
Delle Piane, Paulo: 140
Di Bono, Riccardo: 68
Dini, Spartaco: 93
Donington Park circuit: 108,140,143
Dooley, John: 92,106-107
Dorsay Coupe-de-Ville coachwork: 21-22
Drake Motors, Richard: 4,109,133
Dufty, Les: 180
Driven TV programme: 11
Drovandi, Rinaldo: 106
DTM (Deutsche Tourenwagen
Meisterschaft): 136,141-42

Earls Court Motor Show: 161-62
East African Safari Rally: 68
Eastwood, Joanne: 179
EB Spares, The Italian Connection: 172
Engemann, Liane: 81
European Touring Car Championship: 9-
11,54,66-67,70,74,80-81,95,97-98,
    106
Everton, Richard: 66,170,177

Falco, Carrozzeria: 17
Facetti, Carlo: 67-68
Fagioli, Luigi: 42
Fangio, Juan-Manuel: 37-38,42,47
Farina, Nino: 33,37,42
Farina, Giovanni/Stabilimenti/
Carrozzeria: 33,35,38,40,47,99
Ferguson four-wheel drive system: 132

Ferrari, Scuderia: 21,35,122,164
    250 GT: 38
Fiat: 4,18,114,122,129,135,137,139,
    147,158,176
    Centro Stile: 130,145,152,161
Fiat personnel:
    Da Silva, Walter: 145,151
Fiat vehicles:
    850 Abarth: 9,54
    850 Spider/Coupé: 145
    124 Coupé: 145
    125: 63
    Croma: 114,124-25,127
    Tempra: 130
Fisichella, Giancarlo: 143
Fissore, Rayton: 107
Fitzpatrick, John: 80
Fletcher, Steve: 87
F.L.M. (Federazione Lavoratori
Metalmeccanici): 93
Foden: 21,159
Ford: 122,139
    Capri: 80
    Escort: 80
    GT40: 122
    Ka: 144
    Mondeo: 11,129
    Mustang: 9
Francia, Giorgio: 106,140
Frankfurt Motor Show: 52-54,122

Gache, Phillipe: 143
Galli, Nanni: 9,70,80
Garavini, Carrozzeria Eusebio: 20,22,24
Gardner, Frank: 68
Garrett turbocharger: 105,127,132,159
Geneva Show: 107-108,112,118,127
Ghia, Carrozzeria: 35,36,38,44,47-48
Gilardi, Aquilino: 47
Giovanardi, Fabrizio: 140,164-65
Giudici, Gianni: 106,140,143
Giugiaro, Giorgetto: 6,58,60,70
Glemser, Dieter: 80
Graber, Carrozzeria: 35
Greene, Delmas: 58
Griffin, Steven: 97

Hahne, Hubert: 9,74
Hammond, Brian: 72
Harvey, Mary: 98
Harvey-Bailey, Rhoddy/HB Engineering:
    74,108
Heels, Graham: 121
Helsinki circuit: 143
Hérbert, Jean: 54
Hewland gearbox: 139

Hezemans, Toine: 80-81
Highwood Motor Company: 172
Hill, Charles: 89
Hispano Suiza: 20
Holmes, Archibald: 17
Hockenheim circuit: 143
Honda:
    CBR 600: 113
    650 Bros: 114
Hughes, Donald: 54

Imola circuit: 164-65
International Touring Car Championship
    (ITC): 128,136,141-143
I.R.I.: 26,122
Isotta Fraschini: 14
Italian Job, The: 74,106

Jaeger instruments: 18,24
Jaggard, Paul: 89
Jaguar: 11,122
    S-type: 167
Jolly Club team: 64,66-67,93,170
Just Suds: 4,179

Kay, Roger: 106
Kerridge, Alastair: 155,157
Kershaw, Alwyn: 106
Kirby, Rob: 106,107
KKK Turbocharger: 94-95,98,106,116,
    128
Knockhill circuit: 139
Kwech, Horst: 68

Lafitte, Jacques: 106
Laird, Sam: 106
Lancia: 122,128,135,164
    Aurelia: 44-45
    Dedra: 129
    Delta Integrale: 132,136,139
    Kappa: 169
    Thema: 114,124-25,127
Larini, Nicola: 106,140,143,165
Larrauri, Oscar: 140
Le Mans 24-Hours: 11,28,33,52,122
Lexus:
    IS 2000: 10
Lindsay, Michael: 4,107
Lindstrom, Thomas: 106
Livorno Grand Prix: 30
Lombardi, Francis/Carrozzeria: 47
Longbridge: 48
Lotus: 95
    Cortina: 9-11,54,70-71
    Elise: 161

Elite: 51,152
Lydden Hill circuit: 97,105
Madrid Show: 166
Magneti Marelli:
    Selespeed transmission: 156
    common rail diesel system: 158
Mallory Park circuit: 86,121
Marshall, Alan: 107
Masoero, Fernand: 54,67,69-70
Martini racing team: 143
McDonough, Ed: 4,177-78
Mercedes-Benz: 30
    220: 67
    C-class: 141-42
    E-class: 167
MG:
    MGB: 177
MGS Coachworks: 4,170,174,176
Michelotti, Giovanni: 38,41,47
Milan Motor Show: 26,28
Mildren, Alec: 68
Mille Miglia: 20,21,25,28,30-32,35,38,
    42,46,47,52
Mini/Mini Cooper: 9,48,54-56,74
Minoia, Nando: 25
Modena, Stefano: 140,143
Momo:
    steering wheel: 98,134,138-39,147,
        162
    leather upholstery: 162
Monk, Roger: 4
Monro, David: 86
Mont Ventoux hillclimb: 54
Monviso, Carrozzeria: 35
Monza circuit: 28,62,67-68,164
Morbidelli, Gianni: 106,140
Moretti, Carrozzeria: 56
Morgan, Nathan: 161
Morris, Malcolm: 170
Motor Industry Research Unit: 122
Mozzi, Danilo: 140
Muir, Brian: 80
Mugello circuit: 17
Munaron, Gino: 68
Mussolini, Benito: 17,32

Nanini, Sandro: 106,140,143
Negrete, Lorenzo Quevedo: 59
Nissan: 113
    Cherry: 113-14
Nordauto racing team: 141,165
Nord Italia, Carrozzeria: 21
North, David: 94
North Weald aerodrome: 102
Nuvolari, Tazio: 28,30
Opel:

Calibra: 141
Oreiller, Henri: 54
O.S.I., Carrozzeria: 60-61
Ototrasm transmission system: 117
Oulton Park circuit: 139

Page, Andy: 97
Palmer, Roger: 10
Paris Show: 44
Patrese, Riccardo: 106,128
Perkins diesel engine: 76,78
Pescara 24-Hours: 28
Phillipson, Tony: 102
Pierson engine builder: 81
Pininfarina, Giovan-Battista/Carrozzeria:
    6,28,30,33-35,38,40-42,44,47,
    59,70,118,122
Pinto, Enrico: 68
Porsche:
    356: 51
    911: 147
Portuguese Rally: 68
Presley, Graham: 107,108
Prince Michael of Romania: 33-34
Procar race series: 125,128
Prodrive: 139

Quester, Dieter: 74

RAC Tourist Trophy: 80
Recaro seats: 108,146
Renault:
    R4: 56
    R8: 56
    Laguna: 140
Revelli di Beaumont, Mario: 28
Rice-Edwards, Minette: 105
Ricotti, Count Marco: 13-14,
Riley: 10-11
Rindt, Jochen: 9,10,54,70
Rochebuet, Michel: 66
Roda, Gianluca: 140
Rodriguez, Pedro & Ricardo: 177-78
Roots supercharger: 20
Rossi di Montelera, Theo: 28
Rouse, Andy: 95,98,139
Rover:
    75: 167
Russo, Giacomo 'Geki': 67

Saab:
    9000: 114,124,127
Sala, Carrozzeria: 17
Sala, Luiz Perez: 106
Sandown Park 6-Hours: 68
Satellite navigation system: 167

Savoia Marchetti aircraft: 35
Scaglione, Franco: 47-48
Schlesser, Jean-Louis: 106
Schübel racing team: 143
Schwaben Garage: 111-12
Silverstone circuit: 80,143
Simoni, Giampiero: 136,137,139-40,143
Sims, Monica: 165
Smith, Jonathan: 170
Snetterton circuit: 9,10,54,70,74,87,90,
    97,139
Solana, Moises: 178
Soli, Moreno: 140
Sommer, Raymond: 28
Southern Alfa: 120
Spa Francorchamps circuit: 66-68,81,
    107,141,163
Sparco seat: 138-39
Spenceley, Mike: 4,64,66,68,74,170,173,
    175-76,179-80
Spica fuel injection: 56,71,80-82,92,95-
    96
Squadra Bianca: 69-70
Stewart, Jackie: 9,54,80
Streather, Dave: 114
Stuck, Hans-Joachim: 141
Subaru:
    Legacy: 118
Superturismo race series: 135,137,139-
    41,164-65
Sullivan, Danny: 140
Suspension system: 8,11,17,20,25,29,
    44,49-50,56,64-65,72,75,85,91-92,
    95,97,100,103-104,108-11,114,
    116-17,124-25,128,131-34,136,139,
    142-43,145,147,160,165-67,176
Sweetapple, Chris: 172

Swoboda, Michael: 57
TAG fuel injection system: 139
Tamburini, Antonio: 140-41
Targa Florio: 17,28,38,52
Tarquini, Gabriele: 106,135-39
Taylor, Stuart: 4
Tedeschi, Felice: 140
Thruxton circuit: 92
Titan Industrial Corporation: 56
TOCA: 137
Top Gear TV programme: 164
Tour de France: 46,54
Touring, Carrozzeria: 20-21,25-27,31-39,
    41-42,44,46-47,58-59,99
Tour of Corsica (Tour de Corse): 52
Tour of Italy (Giro d'Italia): 46,54,106
Tour of Sicily: 46
Triumph:
    Dolomite Sprint: 11
Trofeo Garage: 107
Trossi, Count Carlo Felice: 37
Turin Motor Show: 48,50,78,84
TVR: 161
TV Spielfilm racing team: 143
Tyres: 9,20,28,44-45,71,90,109,128,132
    B.F.Goodrich: 109
    Michelin: 9,137,143,165
    Pirelli: 9,20,24
    Toyo: 70
    Yokohama: 109-10,133

Varano circuit: 164
Varzi, Achille: 28,37
Verhage, Eric: 163
Vespa: 10
Vignale, Alfredo/Carrozzeria: 47,56,58
Vidali, Tamara: 140-41

Villamil, Luis: 140
Viotti, Carrozzeria: 6,7,41-42
VM, Stabilimenti Meccanici/turbodiesel:
    94-95,102-103,110,116,118,128,
    132,137
Volkswagen: 176
    Beetle: 147
    Golf: 86-88,141
Volvo: 9
    S80: 167

Warwick, Derek: 139
Watt, Jason: 143
Weymann coachwork: 20-25
Wheels: 9,20,28,38,44,71,90,92,97,
    110,150
    Campagnolo: 67,71,74,87,95,97,172
    Cromodora: 172
    Minilite: 172
    Sankey: 8,17
    Speedline: 132-33,136,139
    O.Z.: 142,165
Whitmore, Sir John: 9-10,54
Wimille, Jean-Pierre: 37
Wing, Colin: 89
Winterbottom, Oliver: 95
Witting da Prato, Vito: 54
Wolleck, Bob: 81

X-Trac sequential shift: 165

Zagato, Elio/Carrozzeria: 8,17-18,20,47,
    52,54,56,60-61,67,80,111-12
Zandvoort circuit: 69-70
ZF gearbox: 69,95,100,167